Lecture Notes in Mathematics

Edited by A. Dold and B. Eckmann

935

Richard Sot

Simple Morphisms
in Algebraic Geometry

Springer-Verlag
Berlin Heidelberg New York 1982

Author
Richard Sot
School of Mathematics, The Institute for Advanced Study
Princeton, NJ 08540, USA

AMS Subject Classifications (1980): 14-XX

ISBN 3-540-11564-1 Springer-Verlag Berlin Heidelberg New York
ISBN 0-387-11564-1 Springer-Verlag New York Heidelberg Berlin

Library of Congress Cataloging in Publication Data
Sot, Richard, 1948- Simple morphisms in algebraic geometry. (Lecture notes in mathematics; 935) Bibliography: p. Includes indexes. 1. Geometry, Algebraic. 2. Morphisms (Mathematics) I. Title. II. Series: Lecture notes in mathematics (Springer-Verlag); 935. QA3.L28 no. 935 [QA564] 510s [516.3'5] 82-10303
ISBN 0-387-11564-1 (U.S.)

This work is subject to copyright. All rights are reserved, whether the whole or part of the material is concerned, specifically those of translation, reprinting, re-use of illustrations, broadcasting, reproduction by photocopying machine or similar means, and storage in data banks. Under § 54 of the German Copyright Law where copies are made for other than private use, a fee is payable to "Verwertungsgesellschaft Wort", Munich.

© by Springer-Verlag Berlin Heidelberg 1982
Printed in Germany

Printing and binding: Beltz Offsetdruck, Hemsbach/Bergstr.
2141/3140-543210

CONTENTS

Chapter 1	The Zariski topology, the Jacobian criterion and examples of simple algebras over a field k	1
Chapter 2	The Kähler 1-differentials	18
Chapter 3	Every k-algebra A which is essentially of finite type over k and simple is a regular local ring	35
Chapter 4	Brief discussion of unramified and étale homomorphisms	45
Chapter 5	Some corollaries to Theorem 3.5	54
Chapter 6	Fitting ideals	57
Chapter 7	Proof of the Jacobian criterion and some characterizations of simple k-algebras and A-algebras	73
Chapter 8	Characterization of simple A-algebras in terms of étale homomorphisms; invariance of the property of being a simple algebra under composition and change of base	89
Chapter 9	Descent of simple homomorphisms and removal of all noetherian assumptions in Chapter 7 and Chapter 8	103
Chapter 10	Simple morphisms of preschemes and translation of previous theorems into the language of preschemes	117
APPENDIX		128
BIBLIOGRAPHY		145
INDEX TO TERMINOLOGY		146
INDEX TO SYMBOLS		146

Supported in part by NSF grant MCS 77-18723 AO4.

CHAPTER 1

The Zariski topology, the Jacobian criterion and examples of simple algebras over a field k

Introduction.

This text treats in detail the concepts of simple algebra over a field k, simple homomorphism of rings, simple algebraic variety, simple morphism of algebraic varieties and simple morphism of preschemes, which all reduce (in our treatment, by definition) to the concept of a simple algebra over a field k.

For the first nine chapters it is only assumed that the reader is acquainted with basic algebra, a few elementary notions in general topology, and a few notions in commutative algebra, the appendix supplying a reference for several of the theorems in commutative algebra cited in the text. In Chapter 10 it is assumed that the reader is well acquainted with the language of preschemes.

For algebraic varieties over the field of complex numbers the concept of simple variety corresponds to the concept of a smooth complex analytic variety. Hence it should not be surprising to observe that the notion of a simple variety is one of the most basic in algebraic geometry. For example, we have the Riemann-Roch theorem for simple projective varieties, the Weil conjectures for a

simple projective variety over a finite field and canonical
classes corresponding to cycles of simple subvarieties of
simple varieties over fields of prime characteristic in
p-adic cohomology.

Given a field k and a finitely generated k-algebra B
we give all of the important characterizations for B to be
simple over k. For example, the Jacobian criterion is
given in Thm. 7.1 (the criterion easiest to apply to
specific examples), the characterization in terms of étale
homomorphisms is given in Thm. 8.2 (one of the easiest
criteria to apply for proving theorems for general B), the
characterization in terms of all of the local rings of B
being geometrically regular is given in Thm. 7.8, and the
characterization in which the Kahler 1-differentials of B
over k are a locally free B-module and the residue class
field extensions over k are separable for all minimal
prime ideals in B is given in Thm. 7.5.

In our treatment we make no assumptions about the field
k. Hence our treatment applies without change to the case
where k is not algebraically closed, where k is not of
characteristic 0 and where k is not perfect. In
addition, we prove the classical Jacobian criterion (Thm.
7.1) by use of fitting ideals (Chapter 6) and demonstrate
its power by applying it in Chapter 1 to determine where
specific curves and surfaces are simple. I learned the
trick of using fitting ideals to prove the Jacobian

criterion from Saul Lubkin. The only difference in my approach being that I do not bother to show that the fitting ideals are independent of the presentation.

The concept of a simple morphism of preschemes (Chapter 10), being local, reduces immediately to the case of a homomorphism of rings. Hence the treatment in the first nine chapters is adequate for most applications in algebraic geometry. Moreover, by treating the concept at the level of a homomorphism of rings in the first nine chapters, there is no need to use sheaves until Chapter 10 where, for convenience, many of the theorems in the first nine chapters are rephrased in the language of preschemes. Thus the material in this text is readily accessible to algebraists as well as to those in other fields with some knowledge of algebra.

The reader will find that Chapter 2 in the text gives a very thorough and readable treatment of the Kahler 1-differentials, leading to the definition and first properties of a simple k-algebra in Chapter 3. The order of presentation of the material in Chapters 3-8 differs from that of many of the standard sources. The text reads well if read in chronological order, the only exception being that the reader may wish to skip Chapter 6 on the first reading and to refer to the Appendix when reading the main body of the text.

Since this is to be only a basic text on simple homomorphisms, we have omitted entirely in our exposition the more general notion of a formally simple homomorphism, the notion of a regularly imbedded closed sub-prescheme of a given prescheme and other characterizations of simple homomorphisms which were not needed to establish the theorems in Chapters 3-9.

1.1. **Definition.** Let A be a ring.
  (1) We denote by Spec(A) the set of prime ideals of the ring A, also called the prime spectrum of A. (We adopt the convention which excludes the ideal A from being prime and which allows the zero ideal to be prime whenever A is not the zero ring.)
  (2) For each $f \in A$ we put $D(f) = \{P \in \text{Spec}(A) | f \notin P\}$. Note that $D(f)$ canonically identifies with $\text{Spec}(A_f)$, and if $f, g \in A$ then $D(f) \cap D(g) = D(fg)$. Hence $\{D(f) | f \in A\}$ is a base for a topology on Spec(A).
  (3) We give Spec(A) the topology with open base $\{D(f) | f \in A\}$, called the Zariski topology on Spec(A).
  (4) Given $P \in \text{Spec}(A)$, P is said to be a generic point of Spec(A) if and only if P is a minimal prime ideal of A, i.e. a minimal element in the set Spec(A).

1.1.1. **Remarks.** Let A be a ring.
  (1) It is readily verified that the Zariski topology on Spec(A) is quasicompact, i.e. every open covering of Spec(A) has a finite subcovering.
  (2) Note that the Zariski topology on Spec(A) is T_0, i.e. given two points in Spec(A), one has an open neighborhood not containing the other.
  (3) Given $P \in \text{Spec}(A)$, $\{P\}$ is a closed set if and only if P is a maximal ideal in A. Hence for a noetherian ring A,

Spec(A) is T_1 if and only if A is artinian. In particular, for a noetherian integral domain A, Spec(A) is T_1 if and only if A is a field.

(4) In view of (2) and (3) we note that the Zariski topology on Spec(A) is rarely Hausdorff, regular or paracompact.

(5) It is readily verified that every nonempty open subset of Spec(A) contains a generic point and that Spec(A) = $\bigcup_g \overline{\{g\}}$, where g runs through the set of generic points of Spec(A).

(6) In view of (5) it is readily verified that if an open subset U of Spec(A) contains every maximal ideal of A, then U = Spec(A).

(7) If A is a finitely generated algebra over an algebraically closed field k, say $A = k[T_1, \ldots, T_N]/(f_1, \ldots, f_M)$, where $N, M \geq 1$ & $f_i \in k[T_1, \ldots, T_N]$ for $1 \leq i \leq M$, then there is a one to one correspondence between the maximal ideals in A and the points $x \in k^N$ satisfying $f_i(x) = 0$ for all $1 \leq i \leq M$ given by $(T_1 - x_1, \ldots, T_N - x_N) \longmapsto (x_1, \ldots, x_N)$. This is a consequence of the Hilbert Nullstellensatz.

See Hartshorne's text on algebraic geometry for further details on Spec(A).

1.2. Notation. Let A be a ring.

(1) For each $P \in \text{Spec}(A)$ we denote by $\kappa(P)$ the residue class field of A_P, which canonically identifies to the quotient

field of A/P.

(2) Given integers $M, N \geq 1$, an $M \times N$ matrix a with entries in A and $P \in \mathrm{Spec}(A)$ we denote by $a(P)$ the matrix obtained from a by replacing each entry of a by its image in $\kappa(P)$ under the canonical homomorphism $A \longrightarrow \kappa(P)$.

1.2.1. Remarks.

(1) Let B be a ring, $Q \in \mathrm{Spec}(B)$ and put $A = B_Q$. Note that $P \longmapsto PA$ gives a canonical one to one correspondence between the prime ideals of B contained in Q and the prime ideals of A. Hence $\mathrm{Spec}(A)$ identifies to a subset of $\mathrm{Spec}(B)$.

(2) With notation as in (1) let $P \in \mathrm{Spec}(B)$ and put $P' = PA$. Note $\kappa(P)$ identifies to $\kappa(P')$.

1.3. Definition. (<u>The Jacobian criterion</u>)

Let k be a field, B be a finitely generated k-algebra, $Q \in \mathrm{Spec}(B)$ and put $A = B_Q$. Since B is a finitely generated k-algebra there exists $N \geq 1$ and an ideal I in $k[T_1, \ldots, T_N]$ such that B identifies to $k[T_1, \ldots, T_N]/I$. Let $M \geq 1$, $\{f_1, \ldots, f_M\}$ be a set of generators for the ideal I and let a denote the matrix $(\partial f_i / \partial T_j)_{1 \leq i \leq M, 1 \leq j \leq N}$. Put $n = \sup_P \mathrm{tr.deg.}\, \kappa(P)/k$, where P runs through the set of minimal prime ideals in A. The following two conditions are equivalent (This is proved in Thm. 7.1 given later.):

(1) $\text{rank}_{\kappa(Q)} \mathfrak{a}(Q) \geq N-n$.

(2) $\text{rank}_{\kappa(Q)} \mathfrak{a}(Q) = N-n$.

(Whether the equivalent conditions (1) and (2) are satisfied is independent of the choice of N, I, M and $\{f_1, \ldots, f_M\}$. This is immediate by Thm. 7.1 given later.)

We say that B <u>is simple over k at Q</u> or that A <u>is simple over k</u> if and only if one of the equivalent conditions (1) or (2) above is satisfied. We say that B <u>is simple over k</u> if and only if B is simple over k at Q for each $Q \in \text{Spec}(B)$.

1.3.1. Remarks.

(1) Def. 1.3 although different from the definition of A being simple over k given later as Def. 3.3, by Thm. 7.1 given later, Def. 1.3 agrees with Def. 3.3. Hence there is no inconsistency here. However, beginning with Chapter 3, we use Def. 3.3 in place of Def. 1.3 (The latter, moreover, needs Thm. 7.1 in order to be well-defined.).

(2) By Thm. A.6 in the appendix we have alternate characterizations of the integer n in Def. 1.3:

(2A) $n = \dim A + \text{tr.deg.}\ \kappa(Q)/k$.

(2B) $n = \sup_{P \in \text{Spec}(A)} \text{tr.deg.}\ \kappa(P)/k$.

In specific examples, however, n is most easily computed in the form of the characterization given in Def. 1.3.

(3) With notation and hypotheses as in Def. 1.3 if B is an integral domain and Q is a maximal ideal of B, then by Thm. A.6 in the appendix $n = \dim B$. Hence if B is a curve (resp., surface) in euclidean N-dimensional space $k[T_1, \ldots, T_N]$ over k, then $n = 1$ (resp., $n = 2$).

(4) With notation and hypotheses as in Def. 1.3, n can differ for two different maximal ideals Q_1, Q_2 of B. For example, take $N = 2$, put $T_1 = X$, $T_2 = Y$, $I = (X(X-1), Y(X-1))$, $Q_1 = (X, Y)$ and $Q_2 = (X-1, Y)$. Then at Q_1, $n = 0$ and at Q_2, $n = 1$. This explains why we needed to assume B is an integral domain in (3). In geometric language (which we shall not explain here) the variety corresponding to B has two irreducible components, one being the line $X = 1$ with generic point $(X-1) \subset Q_2$ and the other being the point Q_1 corresponding to the origin in $k \times k$. Thus our intuition would lead us to guess that $n = 0$ at Q_1 and $n = 1$ at Q_2.

Next we proceed to consider several examples which serve to illustrate the nature of points where B is not simple over k and which illustrate also how easy it is to determine, in practice, where B is simple over k by using the Jacobian criterion.

In examples 1.1-1.9 and exercises 1.1-1.4 below k is a fixed field.

Example 1.1. Let $N \geq 1$ and put $B = k[T_1, \ldots, T_N]$. We proceed to show that B is simple over k. With notation as in Def. 1.3 we have $n = \text{tr.deg.} k(T_1, \ldots, T_N)/k = N$ whence $N-n = 0$. Hence by Def. 1.3 we conclude B is simple over k.

Example 1.2. Every line in $k[X, Y]$ is simple over k. By interchanging X and Y, if necessary, we can assume $B = k[X, Y]/(Y-mX-b)$, where $m, b \in k$ and m and b are not both 0. By Eisenstein's criterion $Y-mX-b$ is irreducible in $k[X, Y]$. Hence B is an integral domain. Let K be the quotient field of B. Hence with notation as in Def. 1.3 we have $n = \text{tr.deg.} K/k = 1$ and $a = (-m, 1)$. Let Q be a prime ideal in $k[X, Y]$ containing $Y-mX-b$. Then $\text{rank}_{\kappa(Q)} a(Q) \neq 0$ since 1 is an entry of a. Hence by Def. 1.3 we conclude B is simple over k.

Example 1.3. Put $B = k[X, Y]/(XY)$. Since (X, Y) corresponds to the origin in $k \times k$ we expect that if $\text{ch}(k) = 0$ and $Q \in \text{Spec}(B)$, then B is simple over k at Q if and only if $Q \neq (X, Y)$. We shall establish this result for all choices of the characteristic.

The images of the ideals (X) and (Y) in $k[X, Y]$ in B are the minimal prime ideals in B. Hence with notation as in Def. 1.3 we have

$$n = \text{tr.deg.}(k[X, Y]/(X))_{(0)}/k = \text{tr.deg.} k(Y)/k = 1$$

and $a = (Y, X)$. Let Q be a prime ideal in $k[X, Y]$ containing XY. Then $\text{rank}_{\kappa(Q)} a(Q) = 0$ if and only if $\{X, Y\} \subset Q$ if and only if

$Q = (X, Y)$. Hence by Def. 1.3 we conclude B is simple over k at Q if and only if $Q \neq (X, Y)$.

Example 1.4. Put $B = k[X, Y]/(2XY)$. If $ch(k) \neq 2$ then $B = k[X, Y]/(XY)$ whence by Example 1.3 we conclude given $Q \in Spec(B)$ that B is simple over k at Q if and only if $Q \neq (X, Y)$. On the other hand, if $ch(k) = 2$ then $B = k[X, Y]$ whence by Example 1.1 we conclude B is simple over k.

Example 1.5. Put $B = k[X, Y]/(X^2+Y^2-1)$. When $ch(k) \neq 2$ and k is algebraically closed the maximal ideals of B correspond to the points on the unit circle in $k \times k$. Hence if $ch(k) = 0$ we expect that B is simple over k.

Suppose first that $ch(k) \neq 2$. By Eisenstein's criterion we conclude X^2+Y^2-1 is irreducible in $k[X, Y]$. Hence B is an integral domain. Let K be the quotient field of B. Hence with notation as in Def. 1.3 we have $n = tr.deg. K/k = 1$ and $\mathfrak{a} = (2X, 2Y) = (X, Y)$. Let Q be a prime ideal in $k[X, Y]$ containing X^2+Y^2-1. Then $\text{rank}_{\kappa(Q)} \mathfrak{a}(Q) = 0$ if and only if $\{X, Y\} \subset Q$ if and only if $Q = (X, Y)$ which is impossible since $\{X^2+Y^2-1, X, Y\} \subset Q$ implies $1 \in Q$. Hence $\text{rank}_{\kappa(Q)} \mathfrak{a}(Q) \geq 1 = n$. Hence by Def. 1.3 we conclude B is simple over k.

Now suppose $ch(k) = 2$. Hence $X^2+Y^2-1 = (X+Y+1)^2$ and $X+Y+1$ is irreducible in $k[X, Y]$. Hence with notation as in Def. 1.3 we have $n = tr.deg.(k[X, Y]/(X+Y+1))_{(0)}/k = 1$ and $\mathfrak{a} = (2X, 2Y) = (0, 0)$. Hence by Def. 1.3 we conclude that for each $Q \in Spec(B)$ that B is not simple over k at Q.

Example 1.6. Put $B = k[X, Y]/(X^3-Y^2)$. When $k = \mathbb{R}$ the points $(x, y) \in k \times k$ corresponding to the maximal ideals of B have the graph of the semicubical parabola in Figure 1. Hence if $ch(k) = 0$ we expect that for each $Q \in Spec(B)$, B is simple over k at Q if and only if $Q \neq (X, Y)$. Computation shows that X^3-Y^2 is irreducible in $k[X, Y]$. Hence B is an integral domain. Let K be the quotient field of B. Hence with notation as in Def. 1.3 we conclude $n = tr.deg. K/k = 1$ and $\mathfrak{a} = (3X^2, -2Y)$. Let $P \in Spec(k[X, Y])$ such that $X^3-Y^2 \in P$ and let Q be the image of P in B. We proceed to show

(1) B is simple over k at Q if and only if $Q \neq (X, Y)$.

Case 1. $ch(k) \notin \{2, 3\}$.

We have $rank_{\kappa(P)} \mathfrak{a}(P) = 0$ if and only if $\{3X^2, -2Y\} \subset P$ if and only if $\{X^2, Y, X^3-Y^2\} \subset P$ if and only if $P = (X, Y)$. Hence by Def. 1.3 we obtain (1).

Case 2. $ch(k) = 2$.

We have $rank_{\kappa(P)} \mathfrak{a}(P) = 0$ if and only if $\{3X^2, -2Y\} \subset P$ if and only if $X^2 \in P$ if and only if $\{X^2, X^3-Y^2\} \subset P$ if and only if $P = (X, Y)$. Hence by Def. 1.3 we obtain (1).

Case 3. $ch(k) = 3$.

This case is similar to Case 2.

Example 1.7. Put $B = k[X, Y]/(Y^2-X^2(X+1))$. When $k = \mathbf{R}$ the points (x, y) in $k \times k$ corresponding to the maximal ideals of B have the graph as illustrated in Figure 2. Hence if $ch(k) = 0$ we expect that for each $Q \in Spec(B)$ that B is simple over k at Q if and only if $Q \neq (X, Y)$.

$Y^2-X^2(X+1)$ is irreducible in $k[X, Y]$ by Eisenstein's criterion. Hence B is an integral domain. Let K be the quotient field of B. Hence with notation as in Def. 1.3 we conclude $n = tr.deg. K/k = 1$ and $\mathfrak{a} = (-3X^2-2X, 2Y)$. Let $P \in Spec(k[X, Y])$ such that $Y^2-X^2(X+1) \in P$ and let Q be the image of P in B. We proceed to show

(1) B is simple over k at Q if and only if $Q \neq (X, Y)$.

We have $rank_{\kappa(Q)} \mathfrak{a}(Q) = 0$ if and only if $\{-3X^2-2X, 2Y\} \subset P$ if and only if $\{-3X^2-2X, 2Y, Y^2-X^2(X+1)\} \subset P$ if and only if $P = (X, Y)$. This establishes (1).

Example 1.8. Put $B = k[X, Y, Z]/(X^2+Y^2-Z^2)$. When $k = \mathbf{R}$ the points (x, y, z) in k^3 corresponding to the maximal ideals of B have the graph as illustrated in Figure 3. Hence if $ch(k) = 0$ we expect that for each $Q \in Spec(B)$, B is simple over k at Q if and only if $Q \neq (X, Y, Z)$.

    Case 1. $ch(k) \neq 2$.

If k does not contain a square root of -1 it is readily verified that X^2+Y^2 is irreducible in $k[X, Y]$ whence by Eisenstein's criterion,

$X^2+Y^2-Z^2$ is irreducible in $k[X, Y, Z]$. Suppose, on the other hand, k contains a square root i of -1. Since $X+iY$ is irreducible in $k[X, Y]$, $(X+iY)^2$ does not divide $(X-iY)$ in $k[X, Y]$ and $X^2+Y^2-Z^2 = (X+iY)(X-iY)-Z^2$, by Eisenstein's criterion we conclude $X^2+Y^2-Z^2$ is irreducible in $k[X, Y, Z]$. Hence in either case B is an integral domain. Let K be the quotient field of B. Hence with notation as in Def. 1.3 we conclude $n = $ tr. deg. $K/k = 2$ whence $N-n = 3-2 = 1$ and $\mathcal{a} = (2X, 2Y, -2Z)$. Let $P \in \mathrm{Spec}(k[X, Y, Z])$ such that $X^2+Y^2-Z^2 \in P$ and let Q be the image of P in B. We proceed to show

(1) B is simple over k at Q if and only if $Q \neq (X, Y, Z)$.

We have $\mathrm{rank}_{\kappa(Q)} \mathcal{a}(Q) = 0$ if and only if $\{2X, 2Y, -2Z\} \subset P$ if and only if $P = (X, Y, Z)$. By Def. 1.3 this establishes (1).

Case 2. $\mathrm{ch}(k) = 2$.
Hence $X^2+Y^2-Z^2 = (X+Y+Z)^2$ and $X+Y+Z$ is irreducible in $k[X, Y, Z]$. Hence with notation as in Def. 1.3 we conclude

$$n = \mathrm{tr.deg.}(k[X, Y, Z]/(X+Y+Z))_{(0)}/k = 2$$

whence $N-n = 3-2 = 1$ and $\mathcal{a} = (2X, 2Y, -2Z) = (0, 0, 0)$. Hence by Def. 1.3 we conclude for each $Q \in \mathrm{Spec}(B)$ that B is not simple over k at Q.

Example 1.9. Let F be a field of characteristic $p \neq 0$. Put $k = F(X)$ and $B = k[S, T]/(T^p-X)$. T^p-X is irreducible in $k[S, T]$

by Eisenstein's criterion. Hence B is an integral domain. Let K be the quotient field of B. Hence with notation as in Def. 1.3 we have $n = \text{tr.deg. } K/k = 1$ and $\mathfrak{a} = (0, pT^{p-1}) = (0, 0)$. Hence by Def. 1.3 we conclude that for each $Q \in \text{Spec}(B)$, B is not simple over k at Q.

Exercises:

1.1. Put $B = k[X, Y]/(Y-X^2)$. (Parabola). Show that B is simple over k.

1.2. Put $B = k[X, Y, Z]/(X^2+Y^2+Z^2-1)$. (Unit sphere). If $\text{ch}(k) \neq 2$ show that B is simple over k. If $\text{ch}(k) = 2$ show that for each $Q \in \text{Spec}(B)$, B is not simple over k at Q.

1.3. Put $B = k[X, Y]/(X^3+Y^3-3XY)$. (Folium of Descartes). If $k = \mathbb{R}$ then the points (x, y) in $k \times k$ corresponding to the maximal ideals of B have the graph as indicated in Figure 4. (Note that if $\text{ch}(k) \neq 3$ direct computation shows that X^3+Y^3-3XY is irreducible in $k[X, Y]$.) If $\text{ch}(k) \neq 3$ show that for each $Q \in \text{Spec}(B)$ that B is simple over k at Q if and only if $Q \neq (X, Y)$. If $\text{ch}(k) = 3$ show that for each $Q \in \text{Spec}(B)$ that B is not simple over k at Q.

1.4. Let $g_2, g_3 \in \mathbb{C}$ such that $\Delta = g_2^3 - 27g_3^2 \neq 0$. Put $B = \mathbb{C}[X, Y]/(Y^2-4X^3+g_2X+g_3)$. (Note B is an elliptic curve over $\mathbb{C}$ and that all elliptic curves over $\mathbb{C}$ are of the form of B.) Since $\Delta \neq 0$, $4X^3-g_2X-g_3$ does not have multiple roots in $\mathbb{C}$ whence $Y^2-4X^3+g_2X+g_3$ is irreducible in $\mathbf{C}[X, Y]$ by Eisenstein's criterion. Hence B is an integral domain. Since $\Delta \neq 0$ it is also readily computed that B is simple over $\mathbf{C}$.

1.3.2. Remarks.

Let $M, N \geq 1$, k be a field, $f_i \in \mathbf{Z}[T_1, \ldots, T_N]$ for $1 \leq i \leq M$, let I be the ideal in $k[T_1, \ldots, T_N]$ generated by $\{f_i | 1 \leq i \leq M\}$ and put $B = k[T_1, \ldots, T_N]/I$. If we fix a characteristic for k, whether B is simple over k is independent of the choice of k: This is Prop. 9.5, a trivial result, and was observed to be the case in Examples 1.1 - 1.9 and Exercises 1.1 - 1.3 above. Moreover, it was observed that in Examples 1.1 - 1.9 and Exercises 1.1 - 1.3 above B is simple over k for $ch(k) = 0$ if and only if B is simple over k for $ch(k) = p$ for all but finitely many prime numbers p. That this is always the case is Prop. 9.6, an elementary result. Moreover, Prop. 9.6 gives a somewhat sharper result, namely that B is simple over k for $ch(k) = 0$ if and only if B is simple over k for $ch(k) = p$ for infinitely many prime numbers p.

That these results cannot be improved is evident by some of the examples and exercises given above. In Example 1.4 we noted that B is simple over k if and only if $ch(k) = 2$. We can sharpen this pathology as follows. Let $\{p_1, \ldots, p_n\}$ be a finite set of prime numbers. Put $B = k[X, Y]/(p_1 \ldots p_n XY)$. Then B is simple over k if and only if $ch(k) = p_i$ for some $1 \leq i \leq n$. On the other hand, in Example 1.5 we noted that B is simple over k if and only if ch(k) ≠ 2. We can sharpen this pathology by modifying B so that B is simple over k if and only if ch(k) ∉ S for various finite sets of prime numbers S. Thus the results mentioned in the above paragraph are best possible.

CHAPTER 2

The Kahler 1-differentials

In this chapter we define and present several basic properties of the Kahler 1-differentials many of which shall be used later without comment. Since this material is very elementary many readers may wish to examine only the definitions and statements of results, postponing the proofs until later.

2.1. Definition. Let A be a ring, B be an A-algebra, let $\pi_{B/A} = \pi_B = \pi : B \otimes_A B \longrightarrow B$ be the epimorphism defined by $\pi(b \otimes c) = bc$ for all $b, c \in B$ and put $I_{B/A} = I_B = I = \text{Ker}(\pi)$. The B-module I/I^2 is denoted $\Gamma_A^1(B)$ and is called the <u>module of Kahler 1-differentials of the A-algebra B</u>. We define a homomorphism of A-modules $d_{B/A} = d : B \longrightarrow \Gamma_A^1(B)$ by $d(b) = \overline{b \otimes 1 - 1 \otimes b}$ where given $x \in I$ we denote the image of x in I/I^2 by $\bar{x}$.

2.1.1. Remark. With notation and hypotheses as in Def. 2.1 many people prefer to write $\Omega_A^1(B)$ or $\Omega_{B/A}^1$ in place of $\Gamma_A^1(B)$.

2.2. Definition. Let A be a ring, B be an A-algebra and M be a B-module. Then an <u>A-derivation of B into M</u> is a homomorphism of A-modules $d : B \longrightarrow M$ such that $d(bc) = bd(c) + cd(b)$ for all $b, c \in B$.

2.2.1. Remark. Let A, B and M be as in Def. 2.2 and let $d : B \longrightarrow M$

be a homomorphism of abelian groups. Then d is an A-derivation of B into M if and only if $d(a) = 0$ for all $a \in A$ and $d(bc) = db(c) + cd(b)$ for all $b, c \in B$. (The verification is trivial.)

2.2.2. Remark. Let A, B and M be as in Def. 2.2. We denote by $\text{Der}_A(B, M)$ the set of A-derivations of B into M. Note that $\text{Der}_A(B, M)$ has the natural structure of an A-module.

2.3. Lemma. Let A be a ring, B be an A-algebra, and I be the ideal in $B \otimes_A B$ defined in Def. 2.1. Then the ideal I is generated by the elements $b \otimes 1 - 1 \otimes b$, where b runs through a set of generators of the A-algebra B.

Proof.

Evidently, $b \otimes 1 - 1 \otimes b \in I$ for all $b \in B$. On the other hand, given $b, c \in B$ we have $b \otimes c = bc \otimes 1 + (b \otimes 1)(1 \otimes c - c \otimes 1)$. If $\sum_i (b_i \otimes c_i) \in I$ we have by definition of I that $\sum_i b_i c_i = 0$ whence $\sum_i (b_i \otimes c_i) = \sum_i (b_i \otimes 1)(1 \otimes c_i - c_i \otimes 1)$ which proves the ideal I is generated by the elements $b \otimes 1 - 1 \otimes b$, $b \in B$. Moreover, if $b, b_1, b_2 \in B$ and $b = b_1 b_2$ we have $b \otimes 1 - 1 \otimes b = (b_1 \otimes 1)(b_2 \otimes 1 - 1 \otimes b_2) + (b_1 \otimes 1 - 1 \otimes b_1)(1 \otimes b_2)$ which completes the proof of the lemma.

2.4. Proposition. Let A be a ring, B be an A-algebra and let $d : B \longrightarrow \Gamma_A^1(B)$ be the homomorphism of A-modules defined in Def. 2.1. The following are true:

(1) The B-module $\Gamma_A^1(B)$ is generated by the elements $d(b)$, where b runs through a set of generators for B as an A-algebra.

(2) d is an A-derivation of B into $\Gamma_A^1(B)$.

(3) The pair $(d, \Gamma_A^1(B))$ is characterized up to a canonical isomorphism by the following universal mapping property: Given a B-module M and an A-derivation e of B into M, there exists a unique homomorphism $f : \Gamma_A^1(B) \longrightarrow M$ of B-modules such that $f \circ d = e$.

(4) For each B-module M, there is a canonical isomorphism of A-modules

$$\Lambda : \text{Hom}_B(\Gamma_A^1(B), M) \longrightarrow \text{Der}_A(B, M) .$$

Proof.

(1) is immediate by Lemma 2.3. Next we proceed to establish (2). Let $a \in A$. Then $d(a) = \overline{a \otimes 1 - 1 \otimes a} = 0$. Let $b, c \in B$. Then $d(bc) = \overline{bc \otimes 1 - 1 \otimes bc} = \overline{bc \otimes 1 - 1 \otimes bc + b \otimes c - b \otimes c} = b(\overline{c \otimes 1 - 1 \otimes c}) + c(\overline{b \otimes 1 - 1 \otimes b}) = bd(c) + cd(b)$. This establishes (2).

We postpone the proof of (3) until after Prop. 2.7 when we shall establish (3) independently of (4).

Finally we proceed to establish (4). Evidently putting $\Lambda(f) = f \circ d$ for all $f \in \text{Hom}_B(\Gamma_A^1(B), M)$ defines a homomorphism of A-modules $\Lambda : \text{Hom}_B(\Gamma_A^1(B), M) \longrightarrow \text{Der}_A(B, M)$ in view of (2). Let $f \in \text{Hom}_B(\Gamma_A^1(B), M)$. Suppose $\Lambda(f) = f \circ d = 0$. Then $f(d(b)) = 0$

for all $b \in B$ which in view of (1) yields $f = 0$. Hence Λ is injective. That Λ is surjective is immediate by applying (3).

2.5. Definition. Let A be a ring, B be an A-algebra and M be a B-module. The B-module $B \times M$ can be given the structure of commutative ring with identity with addition defined componentwise and multiplication defined by $(b_1, m_1)(b_2, m_2) = (b_1 b_2, b_1 m_2 + b_2 m_1)$ for all $b_1, b_2 \in B$ and $m_1, m_2 \in M$. $B \times M$ with this structure shall be denoted by $D_B(M)$.

Propositions 2.6 and 2.7 which appear below shall only be needed to establish condition (4) in Prop. 2.4 and Prop. 2.8 and therefore are only of technical interest here.

2.6. Proposition. Let A be a ring, B be an A-algebra and M be a B-module. Let S be the set of homomorphisms $B \longrightarrow D_B(M)$ of A-algebras such that the composite $B \longrightarrow D_B(M) \longrightarrow B$ is the identity. Then there is a canonical bijection $\Psi : \mathrm{Der}_A(B, M) \longrightarrow S$.

Proof.

Define Ψ by $(\Psi(d))(b) = (b, d(b))$ for all $d \in \mathrm{Der}_A(B, M)$ and $b \in B$. Evidently $\Psi(d) \in S$ for all $d \in \mathrm{Der}_A(B, M)$. It is also evident by construction that Ψ is injective. Hence it remains only to show Ψ is surjective.

For this purpose let $u \in S$. For each $b \in B$ choose $m_b \in M$ such that $u(b) = (b, m_b)$. Define a function $d : B \longrightarrow M$ by

$d(b) = m_b$ for all $b \in B$. d is A-linear since u is A-linear. Now let $b, c \in B$. We have $(bc, m_{bc}) = u(bc) = u(b)u(c) = (b, m_b)(c, m_c) = (bc, bm_c + cm_b)$ whence $m_{bc} = bm_c + cm_b$ whence $d(bc) = bd(c) + cd(b)$. Thus $d \in \mathrm{Der}_A(B, M)$ and by construction $\Psi(c) = u$. Hence Ψ is surjective. This completes the proof of the proposition.

2.7. Proposition. Let A be a ring, B be an A-algebra and I be as in Def. 2.1. Then there is a canonical isomorphism of B-algebras $\Phi : B \otimes_A B / I^2 \longrightarrow D_B(\Gamma_A^1(B))$. Moreover, $\Phi|\Gamma_A^1(B)$ composed with the projection $D_B(\Gamma_A^1(B)) \longrightarrow \Gamma_A^1(B)$ is the identity.

Proof.

Given $b, c \in B$ we put $\Phi(\overline{b \otimes c}) = (bc, \overline{b \otimes c} - \overline{bc \otimes 1})$. It is readily checked that this yields a well-defined homomorphism $\Phi : B \otimes_A B / I^2 \longrightarrow D_B(\Gamma_A^1(B))$ of B-modules. Given $b, c, b', c' \in B$ we have $\Phi((\overline{b \otimes c})(\overline{b' \otimes c'})) = \Phi(\overline{bb' \otimes cc'}) = (bb'cc', \overline{bb' \otimes cc'} - \overline{bb'cc' \otimes 1}) = (bcb'c', \overline{bb' \otimes cc'} - \overline{bb'c \otimes c'} + \overline{bcb' \otimes c'} - \overline{bcb'c' \otimes 1}) = (bcb'c', b'c'(\overline{b \otimes c} - \overline{bc \otimes 1}) + bc(\overline{b' \otimes c'} - \overline{b'c' \otimes 1})) =$
$= (bc, \overline{b \otimes c} - \overline{bc \otimes 1})(b'c', \overline{b' \otimes c'} - \overline{b'c' \otimes 1}) = \Phi(\overline{b \otimes c})\Phi(\overline{b' \otimes c'})$ and $\Phi(\overline{1 \otimes 1}) = (1, 0)$ whence Φ is a homomorphism of B-algebras.

Now let $\sum_i b_i \otimes c_i \in I$, i.e. $b_i, c_i \in B$ for all i and $\sum_i b_i c_i = 0$, i.e. $\sum_i \overline{(b_i \otimes c_i)} \in \Gamma_A^1(B)$. Then $\Phi(\sum_i (b_i \otimes c_i)) = \sum_i \Phi(\overline{b_i \otimes c_i}) = \sum_i (b_i c_i, \overline{b_i \otimes c_i} - \overline{b_i c_i \otimes 1}) = (\sum_i b_i c_i, \sum_i \overline{b_i \otimes c_i} - \sum_i \overline{(b_i c_i \otimes 1)}) = (0, \sum_i \overline{b_i \otimes c_i})$.

Hence $\Gamma_A^1(B) \xrightarrow{\Phi|\Gamma_A^1(B)} D_B(\Gamma_A^1(B)) \xrightarrow{\text{proj.}} \Gamma_A^1(B)$ is the identity. Thus it remains only to show that Φ is bijective.

Define a homomorphism of B-modules $\Phi_0 : D_B(\Gamma_A^1(B)) \longrightarrow B \otimes_A B/I^2$ by $\Phi_0(b_1, b_2 d(b_3)) = \overline{b_1 \otimes 1} + b_2 d(b_3)$ for all $b_1, b_2, b_3 \in B$. It is readily checked that both $\Phi \circ \Phi_0$ and $\Phi_0 \circ \Phi$ are the identity maps. Hence Φ is indeed bijective. This completes the proof of the proposition.

Proof of condition (3) in Prop. 2.4:

In view of Prop. 2.6, Prop. 2.7 and the construction of Ψ and Φ it suffices to prove that each $u \in S$ (S as in Prop. 2.6) factors $u = v \circ p_2$, where $p_2 : B \longrightarrow B \otimes_A B/I^2$ is the homomorphism defined by $p_2(b) = \overline{1 \otimes b}$ for $b \in B$ and such that $v : B \otimes_A B/I^2 \longrightarrow D_B(M)$ is a homomorphism of B-algebras. Let $p_1 : B \longrightarrow B \otimes_A B/I^2$ be the homomorphism defined by $p_1(b) = \overline{b \otimes 1}$ for $b \in B$. Define homomorphisms $j_1, j_2 : B \longrightarrow B \otimes_A B$ analogous to p_1 and p_2 and a homomorphism of B-algebras $j : B \longrightarrow D_B(M)$ by $j(b) = (b, 0)$ for $b \in B$. Since $j(a) = u(a)$ for all $a \in A$ We conclude that there exists a unique homomorphism of A-algebras $w : B \otimes_A B \longrightarrow D_B(M)$ such that both triangles in the diagram commute.

By definition we have

$w(b \otimes 1 - 1 \otimes b) = j(b) - u(b) \in \{0\} \times M$

for all $b \in B$. Hence by Lemma 2.3 this implies $w(I) \subset \{0\} \times M$ whence $w(I^2) = 0$. Hence w factors $B \otimes_A B \longrightarrow B \otimes_A B/I^2 \xrightarrow{v} D_B(M)$, where v is a homomorphism of A-algebras. Moreover, since $v \circ p_1 = j$ is a homomorphism of B-algebras, it is thus the case for v by definition of the structure of the B-algebra $B \otimes_A B / I^2$. Since by definition we have $u = v \circ p_2$, this completes the proof of condition (3) in Prop. 2.4.

2.8. Proposition. Let A be a ring, I be a set, T_i be an indeterminate for each $i \in I$ and put $B = A[(T_i)_{i \in I}]$. Then $\Gamma_A^1(B)$ is a free B-module with basis $\{d(T_i) | i \in I\}$.

Proof.

By (1) in Prop. 2.4 $\{dT_i | i \in I\}$ generates the B-module $\Gamma_A^1(B)$. Now let M be a free B-module with basis $\{m_i | i \in I\}$. Define a homomorphism of A-algebras $u : B \longrightarrow D_B(M)$ by $u(T_i) = (T_i, m_i)$ for all $i \in I$. Hence by Prop. 2.6 we conclude $e = \Psi^{-1}(0) \in \text{Der}_A(B, M)$ satisfies $e(T_i) = m_i$ for all $i \in I$. Hence by (3) in Prop. 2.4 we conclude that there exists a unique homomorphism of B-modules $f : \Gamma_A^1(B) \longrightarrow M$ such that $f \circ d = e$. Thus if $\{d(T_i) | i \in I\}$ were linearly dependent in $\Gamma_A^1(B)$ we would have $\{f(d(T_i)) | i \in I\} = \{e(T_i) | i \in I\} = \{m_i | i \in I\}$ would be linearly dependent in M, a contradiction. This completes the proof of the proposition.

2.9. Proposition. Let A be a ring, B be an A-algebra and C be a B-algebra. Then we have an exact sequence of C-modules

$$\Gamma_A^1(B) \otimes_B C \longrightarrow \Gamma_A^1(C) \longrightarrow \Gamma_B^1(C) \longrightarrow 0 .$$

Proof.

Let $\pi_B : B \otimes_A B \longrightarrow B$ and $\pi_C : C \otimes_A C \longrightarrow C$ be the epimorphisms defined in Def. 2.1 with kernels I_B and I_C. Since I_B maps into I_C under the homomorphism of rings $B \otimes_A B \longrightarrow C \otimes_A C$ we have an induced homomorphism of rings $B \otimes_A B / I_B^2 \longrightarrow C \otimes_A C / I_C^2$ which when restricted to $\Gamma_A^1(B)$ yields a homomorphism $\Gamma_A^1(B) \longrightarrow \Gamma_A^1(C)$ of abelian groups which is $B \longrightarrow C$ linear whence the induced homomorphism $\Gamma_A^1(B) \otimes_B C \longrightarrow \Gamma_A^1(C)$ is C-linear. An argument analogous to the preceding yields a homomorphism of C-modules $\Gamma_A^1(C) \longrightarrow \Gamma_B^1(C)$. This latter homomorphism is surjective since $I_{C/A}$ maps onto $I_{C/B}$ under the homomorphism $C \otimes_A C \longrightarrow C \otimes_B C$. Hence it remains only to show $\text{Im}(\Gamma_A^1(B) \otimes_B C \longrightarrow \Gamma_A^1(C)) = \text{Ker}(\Gamma_A^1(C) \longrightarrow \Gamma_B^1(C))$.

For this purpose it suffices to show that the sequence

(1) $\quad 0 \longrightarrow \text{Hom}_C(\Gamma_B^1(C), M) \longrightarrow \text{Hom}_C(\Gamma_A^1(C), M) \longrightarrow \text{Hom}_C(\Gamma_A^1(B) \otimes_B C, M)$

is exact at $\text{Hom}_C(\Gamma_A^1(C), M)$ for each C-module M. But in view of condition (4) in Prop. 2.4 the sequence (1) becomes

(2) $\quad 0 \longrightarrow \text{Der}_B(C, M) \xrightarrow{\lambda} \text{Der}_A(C, M) \xrightarrow{\mu} \text{Hom}_C(\Gamma_A^1(B) \otimes_B C, M)$.

Hence it suffices to show that the sequence (2) is exact at $\text{Der}_A(C, M)$. First we proceed to show that $\mu \circ \lambda = 0$. Let $e \in \text{Der}_B(C, M)$. By condition (3) in Prop. 2.4 there exists a unique homomorphism of B-modules $f : \Gamma_A^1(B) \longrightarrow M$ such that $f \circ d_{B/A} = e \circ \beta$, where $\beta : B \longrightarrow C$ is the structure homomorphism. Hence given $b \in B$ and $c \in C$ we have
$(\mu \circ \lambda)(e)(d(b) \otimes c) = cf(d(b)) = ce(b) = 0$ since e is a B-derivation whence in view of condition (1) in Prop. 2.4 we conclude $\mu \circ \lambda = 0$.

Now let $e \in \text{Der}_A(C, M)$ such that $\mu(e) = 0$. Let f be as above. Hence $0 = \mu(e)(d(b) \otimes 1) = f(d(b)) = e(b)$ for all $b \in B$ whence $e \in \text{Der}_B(C, M)$. Thus $\text{Ker}(\mu) \subset \text{Im}(\lambda)$. This proves (2) is exact at $\text{Der}_A(C, M)$ and establishes the proposition.

2.10. Proposition. Let A be a ring, B be an A-algebra, J be an ideal in B and put $C = B/J$. Then we have an exact sequence of C-modules

$$J/J^2 \longrightarrow \Gamma_A^1(B) \underset{B}{\otimes} C \longrightarrow \Gamma_A^1(C) \longrightarrow 0 \ .$$
$$\| $$
$$\Gamma_A^1(B)/J \cdot \Gamma_A^1(B)$$

Proof.

By condition (1) in Prop. 2.4 we conclude $\Gamma_B^1(C) = 0$. Hence applying Prop. 2.9 we obtain an exact sequence of C-modules

(1) $\qquad \Gamma_A^1(B) \underset{B}{\otimes} C \xrightarrow{\mu} \Gamma_A^1(C) \longrightarrow 0 \ .$

Put $I = I_{B/A}$ and $d = d_{B/A}$. Then the composite

(2) $\qquad J \longrightarrow B \xrightarrow{d} I/I^2 = \Gamma_A^1(B)$

is a homomorphism of B-modules. Let $b \in J^2$. Then $b = \sum_i b_i c_i$, $b_i, c_i \in J$. Hence $b \otimes 1 - 1 \otimes b = \sum_i b_i(c_i \otimes 1 - 1 \otimes c_i) \in I \cap (J \cdot B \otimes_A B) \subset I^2 + (I \cap J \cdot B \otimes_A B)$. Thus $d(J) \subset I^2 + (I \cap J \cdot B \otimes_A B)$. Hence the composite (2) induces a homomorphism of C-modules

$\lambda : J/J^2 \longrightarrow I/(I^2 + (I \cap J \cdot B \otimes_A B)) = \Gamma_A^1(B)/J \cdot \Gamma_A^1(B)$.

In view of (1) it thus suffices to show $\text{Im}(\lambda) = \text{Ker}(\mu)$. Hence it suffices to show that for each C-module M, the sequence of homomorphisms of C-modules

(3) $\quad 0 \longrightarrow \text{Hom}_C(\Gamma_A^1(C), M) \longrightarrow \text{Hom}_C(\Gamma_A^1(B) \otimes_B C, M) \longrightarrow$

$\longrightarrow \text{Hom}_C(J/J^2, M)$

is exact at $\text{Hom}_C(\Gamma_A^1(B) \otimes_B C, M)$. But in view of condition (4) in Prop. 2.4 the sequence (3) is

(4) $\quad 0 \longrightarrow \text{Der}_A(C, M) \xrightarrow{\mu'} \text{Der}_A(B, M) \xrightarrow{\lambda'} \text{Hom}_C(J/J^2, M)$.

Hence it suffices to show $\text{Im}(\mu') = \text{Ker}(\lambda')$.

For this purpose let $e \in \text{Der}_A(C, M)$ and let $\beta : B \longrightarrow C$ be the canonical epimorphism. By condition (3) in Prop. 2.4 there exists a unique homomorphism of B-modules $f : \Gamma_A^1(B) \longrightarrow M$ such that $f \circ d_{B/A} = e \circ \beta$. Let $b \in J$. We have $f(d_{B/A}(b)) = e(b) = 0$. Thus in view of condition (1) in Prop. 2.4 and the fact that f is B-linear we conclude $\lambda' \circ \mu' = 0$.

Now let $e \in \text{Ker}(\lambda')$. By condition (3) in Prop. 2.4 there exists a unique homomorphism of B-modules $f : \Gamma_A^1(B) \longrightarrow M$ such that $f \circ d_{B/A} = e$. Since $e \in \text{Ker}(\lambda')$ we conclude $f(d_{B/A}(b)) = 0$ for all $b \in J$. Hence $e(b) = 0$ for all $b \in J$, whence e factors $e = e' \circ \beta$ where $e' \in \text{Der}_A(C, M)$. Hence $e = \mu'(e')$. This shows that $\text{Im}(\mu') = \text{Ker}(\lambda')$ and thus completes the proof of the proposition.

2.10.1. **Corollary.** Let A be a ring, I be a set, T_i be an indeterminate for each $i \in I$, put $B = A[(T_i)_{i \in I}]$, let J be an ideal in B and put $C = B/J$. The following are true:

(1) $\Gamma_A^1(C)$ canonically identifies to

$$\bigoplus_{i \in I} Bd(T_i)/(\sum_{i \in I} Jd(T_i) + d(J)B) \text{ and } \Gamma_A^1(B)$$

canonically identifies to $\bigoplus_{i \in I} Bd(T_i)$, where $d = d_{B/A}$.

(2) $d_{C/A}$ is obtained from d by passing to quotients, that is the diagram below commutes:

$$\begin{array}{ccc} B & \xrightarrow{d} & \bigoplus_{i \in I} Bd(T_i) \\ \downarrow & & \downarrow \\ C & \xrightarrow{d_{C/A}} & \bigoplus_{i \in I} Bd(T_i)/(\sum_{i \in I} Jd(T_i) + d(J)B) \end{array}$$

Proof.

Immediate by Prop. 2.10, Prop. 2.8 and condition (3) of Prop 2.4.

2.10.2. **Corollary.** Let A be a ring and C be a finitely presented A-algebra. Then $\Gamma_A^1(C)$ is a finitely presented C-module. In particular, choosing $n \geq 1$ and a finitely generated ideal J in $A[T_1, \ldots, T_n]$ such that C identifies to $A[T_1, \ldots, T_n]/J$ and a set of generators $\{f_1, \ldots, f_m\}$ for the ideal J, letting $\lambda : C^m \longrightarrow C^n$ be the homomorphism of free C-modules corresponding to the matrix $(\partial f_j / \partial T_i)_{1 \leq j \leq m, 1 \leq i \leq n}$ and $\mu : C^n \longrightarrow \Gamma_A^1(C)$ be the homomorphism of C-modules defined by $\mu(c_1, \ldots, c_n) = \sum_{i=1}^{n} c_i d(T_i)$,

the sequence of homomorphisms of C-modules $C^m \xrightarrow{\lambda} C^n \xrightarrow{\mu} \Gamma_A^1(C) \longrightarrow 0$
is exact and gives therefore a presentation for $\Gamma_A^1(C)$ as a C-module.

Proof.

By condition (1) in Cor. 2.10.1 we have

(1) $\qquad \Gamma_A^1(C) = \bigoplus_{i=1}^{n} Bd(T_i) / (\sum_{i=1}^{n} Jd(T_i) + d(J)B)$.

By condition (1) in Prop. 2.4 we conclude that μ is surjective. Moreover, given $c = (c_1, \ldots, c_m) \in C^m$ we have

$$\mu(\lambda(c)) = \mu((\sum_{j=1}^{m} c_j (\partial f_j / \partial T_1), \ldots, \sum_{j=1}^{m} c_j (\partial f_j / \partial T_n))) =$$

$$= \sum_{i=1}^{n} (\sum_{j=1}^{m} c_j (\partial f_j / \partial T_i)) d(T_i) = \sum_{j=1}^{m} (c_j \sum_{i=1}^{n} (\partial f_j / \partial T_i) d(T_i))$$

$$= \sum_{j=1}^{m} c_j d(f_j) = 0 \text{ in view of (1).}$$

Hence it remains only to show $\text{Ker}(\mu) \subset \text{Im}(\lambda)$. For this purpose let $c = (c_1, \ldots, c_n) \in \text{Ker}(\mu)$. Then

(2) $\qquad 0 = \mu(c) = \sum_{i=1}^{n} c_i d(T_i)$.

For $1 \leq i \leq n$ choose $g_i \in B$ such that $\beta(g_i) = c_i$, where $\beta : B \longrightarrow C$ is the canonical epimorphism. By (1) and (2) we conclude $g = \sum_{i=1}^{n} g_i d(T_i) \in \sum_{i=1}^{n} Jd(T_i) + d(J)B$. Hence there exist $h \in J$ and $b \in B$ such that $g - bd(h) \in \sum_{i=1}^{n} Jd(T_i)$. For $1 \leq j \leq m$ choose

$b_j \in B$ such that $h = \sum_{j=1}^{m} f_j b_j$. Then

$$g - \sum_{j=1}^{m} bb_j (\sum_{i=1}^{n} (\partial f_j / \partial T_i) d(T_i)) \in \sum_{i=1}^{n} Jd(T_i).$$

Hence $g_i d(T_i) - \sum_{j=1}^{m} bb_j (\partial f_j / \partial T_i) d(T_i) \in Jd(T_i)$ for $1 \leq i \leq n$, whence $g_i - \sum_{j=1}^{m} bb_j (\partial f_j / \partial T_i) \in J$ for $1 \leq i \leq n$. Hence $\lambda((\beta(bb_i), \ldots, \beta(bb_m))) = c$. This completes the proof of the corollary.

2.10.3. Corollary. Let A, B, I, J and C be as in Cor. 2.10.1. Let K be a set and put $C' = C[(S_k)_{k \in K}]$. Then we have an exact sequence

$$0 \longrightarrow \Gamma_A^1(C) \underset{C}{\otimes} C' \longrightarrow \Gamma_A^1(C') \longrightarrow \Gamma_C^1(C') \longrightarrow 0$$

of homomorphisms of C'-modules.

Proof.

By Prop. 2.9 we have an exact sequence

(1) $$\Gamma_A^1(C) \underset{C}{\otimes} C' \longrightarrow \Gamma_A^1(C') \longrightarrow \Gamma_C^1(C') \longrightarrow 0$$

of homomorphisms of C-modules. Put $B' = B[(S_k)_{k \in K}]$, $J' = JB'$ and let $\lambda : \Gamma_A^1(C) \underset{C}{\otimes} C' \longrightarrow \Gamma_A^1(C')$ be the homomorphism in the sequence (1).

But by Cor. 2.10.1 $\Gamma_A^1(C) \underset{C}{\otimes} C'$ identifies to

$(\underset{i \in I}{\oplus} Bd(T_i) / (\sum_{i \in I} Jd(T_i) + d(J)B)) \underset{C}{\otimes} C' \equiv M$ and $\Gamma_A^1(C')$ identifies to

$((\underset{i \in I}{\oplus} B'd(T_i)) \oplus (\underset{k \in K}{\oplus} B'd(S_k))) / (\sum_{i \in I} J'd(T_i) + \sum_{k \in K} J'd(S_k) + d(J')B') \equiv N$

and λ identifies to the canonical homomorphism $M \longrightarrow N$ which is evidently injective. This establishes the corollary.

2.11. Proposition. Let A be a ring and B and C be A-algebras. Then the canonical homomorphism $\Gamma_A^1(B) \otimes_A C \longrightarrow \Gamma_C^1(B \otimes_A C)$ is bijective.

Proof.

We proceed to show that the pair $(d_{B/A} \otimes_A C, \Gamma_A^1(B) \otimes_A C)$ satisfies the universal mapping property characterization (condition (3) of Prop. 2.4) for $\Gamma_C^1(B \otimes_A C)$. This will complete the proof. Since $d_{B/A} \in \text{Der}_A(B, \Gamma_A^1(B))$ we obtain $d_{B/A} \otimes_A C \in \text{Der}_C(B \otimes_A C, \Gamma_A^1(B) \otimes_A C)$. Now let M be a $B \otimes_A C$-module and $e \in \text{Der}_C(B \otimes_A C, M)$. It remains only to show there exists a unique homomorphism $g : \Gamma_A^1(B) \otimes_A C \longrightarrow M$ of $B \otimes_A C$-modules such that

(1) $$g \circ (d_{B/A} \otimes_A C) = e \ .$$

Let $j : B \longrightarrow B \otimes_A C$ be the homomorphism $b \longmapsto b \otimes 1$. Since $e \circ j \in \text{Der}_A(B, M)$, by the universal mapping property characterization (condition (3) of Prop. 2.4) for $\Gamma_A^1(B)$ we conclude that there exists a unique homomorphism $f : \Gamma_A^1(B) \longrightarrow M$ of B-modules such that

(2) $$f \circ d_{B/A} = e \circ j \ .$$

Define a homomorphism $g : \Gamma_A^1(B) \otimes_A C \longrightarrow M$ of $B \otimes_A C$-modules by $g(x \otimes c) = cf(x)$ for all $x \in \Gamma_A^1(B)$ and all $c \in C$. In view of (2) we obtain (1) and by construction it is readily verified that g is uniquely determined.

2.12.1. Lemma. Let A be a ring, B be an A-algebra with structure homomorphism $\lambda : A \longrightarrow B$, let S be a multiplicatively closed subset of A and suppose

(1) $\lambda(S)$ is a subset of the invertible elements of B.

Then the canonical homomorphism $\Gamma_A^1(B) \longrightarrow \Gamma_{S^{-1}(A)}^1(B)$ is bijective.

Proof.

We proceed to show that the pair $(d, \Gamma_{S^{-1}(A)}^1(B))$ where $d = d_{B/S^{-1}(A)}$ satisfies the universal mapping property characterization (condition (3) of Prop. 2.4) for $\Gamma_A^1(B)$. This will complete the proof.

Evidently $d \in \mathrm{Der}_{S^{-1}(A)}(B, \Gamma_{S^{-1}(A)}^1(B)) \subset \mathrm{Der}_A(B, \Gamma_{S^{-1}(A)}^1(B))$.
Now let M be a B-module and $e \in \mathrm{Der}_A(B, M)$. It remains only to show that there exists a unique homomorphism of B-modules $f : \Gamma_{S^{-1}(A)}^1(B) \longrightarrow M$ such that $f \circ d = e$. Let $s \in S$. We have $0 = e(1) = e(s \cdot (1/s)) = se(1/s)$ since $e \in \mathrm{Der}_A(B, M)$ whence $e(1/s) = 0$ since $\lambda(s)$ is an invertible element of B in view of (1). Hence $e \in \mathrm{Der}_{S^{-1}(A)}(B, M)$. Hence in view of the universal mapping property characterization (condition (3) of Prop. 2.4) for $\Gamma_{S^{-1}(A)}^1(B)$ we conclude that there exists a unique homomorphism of B-modules $f : \Gamma_{S^{-1}(A)}^1(B) \longrightarrow M$ such that $f \circ d = e$. This completes the proof.

2.12.2. Lemma. Let A be a ring, B be an A-algebra and let T be a multiplicatively closed subset of B. Then the canonical homomorphism $\Gamma_A^1(T^{-1}(B)) \longrightarrow T^{-1}(\Gamma_A^1(B))$ is bijective.

Proof.

Put $d = d_{B/A}$ and let $j_1 : B \longrightarrow T^{-1}(B)$ and $j_2 : \Gamma^1_A(B) \longrightarrow T^{-1}(\Gamma^1_A(B))$ be the canonical homomorphisms. Define a function $d' : T^{-1}(B) \longrightarrow T^{-1}(\Gamma^1_A(B))$ by $d'(b/t) = (td(b) - bd(t))/t^2$ for all $b \in B$ and $t \in T$. Then $d' \in \text{Der}_A(T^{-1}(B), T^{-1}(\Gamma^1_A(B)))$ and $d' \circ j_1 = j_2 \circ d$.

We proceed to show that the pair $(d', T^{-1}(\Gamma^1_A(B)))$ satisfies the universal mapping property characterization (condition (3) of Prop. 2.4) for $\Gamma^1_A(T^{-1}(B))$. Let M be a $T^{-1}(B)$-module and $e \in \text{Der}_A(T^{-1}(B), M)$. It remains only to show that there exists a unique homomorphism $f' : T^{-1}(\Gamma^1_A(B)) \longrightarrow M$ of $T^{-1}(B)$-modules such that $f' \circ d' = e$. Note that $e \circ j_1 \in \text{Der}_A(B, M)$. Hence by the universal mapping property characterization (condition (3) of Prop. 2.4) for $\Gamma^1_A(B)$ there exists a unique homomorphism $f : \Gamma^1_A(B) \longrightarrow M$ of B-modules such that $f \circ d = e \circ j_1$. Let f' be the homomorphism $T^{-1}(\Gamma^1_A(B)) \longrightarrow M$ obtained from f by passing to quotients. Let $b \in B$ and $t \in T$. Then $f'(d'(b/t)) =$
$= f'((td(b) - bd(t))/t^2) = (f(td(b) - bd(t)))/t^2 = (tf(d(b)) - bf(d(t)))/t^2 =$
$= (te(b) - be(t))/t^2 = e(b/t)$. Hence $f' \circ d' = e$. It remains only to show that f' is uniquely determined.

For this purpose let $g_1, g_2 : T^{-1}(\Gamma^1_A(B)) \longrightarrow M$ be two homomorphisms of $T^{-1}(B)$-modules such that $g_1 \circ d' = e = g_2 \circ d'$. Put $g = g_1 - g_2$. It suffices to show $g = 0$. Note $g \circ d' = 0$. Let $x \in \Gamma^1_A(B)$ and $t \in T$. By condition (1) in Prop. 2.4 there exists $n \geq 1$ and $\{b_1, \ldots, b_n, c_1, \ldots, c_n\} \subset B$ such that $x = b_1 d(c_1) + \ldots + b_n d(c_n)$ whence $x/t = (b_1/t)d(c_1) + \ldots + (b_n/t)d(c_n)$. Hence $g(x/t) = (b_1/t)g(d(c_1)) + \ldots + (b_n/t)g(d(c_n)) = 0$, since $g \circ d' = 0$.

Thus $g = 0$, completing the proof.

2.13. Proposition. Let A be a ring, B be an A-algebra with structure homomorphism $\lambda : A \longrightarrow B$, let S be a multiplicatively closed subset of A and T be a multiplicatively closed subset of B such that $\lambda(S) \subset T$. Then we have canonical isomorphisms

$$\Gamma^1_{S^{-1}(A)}(T^{-1}(B)) \simeq \Gamma^1_A(T^{-1}(B)) \simeq T^{-1}(\Gamma^1_A(B))$$

of $T^{-1}(B)$-modules.

Proof.

Immediate by Lemma 2.12.1 and Lemma 2.12.2.

2.13.1. Corollary. Let A be a ring, B be an A-algebra with structure homomorphism $\lambda : A \longrightarrow B$, $Q \in \text{Spec}(B)$ and put $P = \lambda^{-1}(Q)$. Then we have canonical isomorphisms

$$\Gamma^1_{A_P}(B_Q) \simeq \Gamma^1_A(B_Q) \simeq \Gamma^1_A(B)_Q$$

of B_Q-modules.

Proof.

Immediate by Prop. 2.13 with $S = A \backslash P$ and $T = B \backslash Q$.

CHAPTER 3

<u>Every k-algebra A which is essentially of</u>
<u>finite type over k and simple is a</u>
<u>regular local ring</u>

In this chapter we proceed to prove that every k-algebra A which is the localization of a finitely generated k-algebra at a prime ideal and simple is a regular local ring, where k is any field. This is Theorem 3.5. The notion of simple k-algebra is presented in Definition 3.3. It will not be until Theorem 7.1 in Chapter 7 that we shall prove that Definition 3.3 agrees with Definition 1.3, the Jacobian criterion, introduced in Chapter 1. It will be apparent in later chapters that Theorem 3.5 has far reaching consequences. We begin first with Proposition 3.1 which strengthens Proposition 2.10 when certain additional hypotheses are satisfied.

3.1. Proposition. Let A be a local ring containing a field k, let m denote the maximal ideal of A and put $K = A/m$. Suppose that K is a separable field extension of k. Then we have an exact sequence of K-vector spaces

(1) $\qquad 0 \longrightarrow m/m^2 \overset{\delta}{\longrightarrow} \Gamma^1_k(A) \underset{A}{\otimes} K \longrightarrow \Gamma^1_k(K) \longrightarrow 0$.

Proof.

By Prop. 2.10 the sequence (1) is right exact. Hence it remains only to show that δ is injective. Put $B = A/m^2$. Then B is a local ring with maximal ideal m/m^2, B contains k, $(m/m^2)^2 = (0)$ and $B/(m/m^2) = K$. Hence applying Prop. 2.10 to $k \to B \to K$ yields an exact sequence of K-vector spaces $m/m^2 \to \Gamma^1_k(B) \otimes_B K \to \Gamma^1_k(K) \to 0$. Since we have a commutative diagram of homomorphisms of K-vector spaces (diagram (2)), the induced map on the kernels of the horizontal maps in diagram (1) yields the commutative diagram (3). Hence to show that $\delta = \delta_A$ is injective it suffices to show that δ_B is injective. By replacing A by B we may thus assume without loss of generality that $m^2 = (0)$. Hence

$$\begin{array}{ccc} \Gamma^1_k(A) \otimes_A K & \to & \Gamma^1_k(K) \\ \downarrow & (2) & \| \\ \Gamma^1_k(B) \otimes_B K & \to & \Gamma^1_k(K) \end{array}$$

$$\begin{array}{ccc} m/m^2 & \xrightarrow{\delta = \delta_A} & \Gamma^1_k(A) \otimes_A K \\ \| & (3) & \downarrow \\ m/m^2 & \xrightarrow{\delta_B} & \Gamma^1_k(B) \otimes_B K \end{array}$$

by Thm. A.1 (in the appendix) we conclude that A contains a field L mapped isomorphically onto K by the canonical homomorphism $A \to K$. As above, the commutative diagram (4) of homomorphisms of L-vector spaces yields the commutative diagram (5) of homomorphisms of L-vector spaces. Hence to show δ is injective it suffices to show δ_0 is injective. By replacing k by L we may thus assume

$$\begin{array}{ccc} \Gamma^1_k(A) \otimes_A K & \to & \Gamma^1_k(K) \to 0 \\ \downarrow & (4) & \downarrow \\ \Gamma^1_L(A) \otimes_A K & \to & \Gamma^1_L(K) \to 0 \end{array}$$

$$\begin{array}{ccc} m/m^2 & \xrightarrow{\delta} & \Gamma^1_k(A) \otimes_A K \\ \| & (5) & \downarrow \\ m/m^2 & \xrightarrow{\delta_0} & \Gamma^1_L(A) \otimes_A K \end{array}$$

without loss of generality that k is mapped onto K by the canonical homomorphism $A \longrightarrow K$.

To show δ is injective it suffices to show that the homomorphism

$$\delta' : \text{Hom}_k(\Gamma_k^1(A) \otimes_A k, k) \longrightarrow \text{Hom}_k(\mathfrak{m}, k)$$

of dual vector spaces is surjective. By Prop. 2.4 we have isomorphisms of k-vector spaces $\text{Hom}_k(\Gamma_k^1(A) \otimes_A k, k) \simeq \text{Hom}_A(\Gamma_k^1(A), k) \simeq \text{Der}_k(A, k)$. If $d \in \text{Der}_k(A, k)$, then $\delta'(d) = d|\mathfrak{m}$. Let $h \in \text{Hom}_k(\mathfrak{m}, k)$. For any $a \in A$ we can write $a = b+c$, $b \in k$, $c \in \mathfrak{m}$ in a unique way. Define $d(a) = h(c)$. $d : A \longrightarrow k$ is evidently additive and by construction $d(b) = 0$ for all $b \in k$. Moreover, letting $a_1 = b_1+c_1$ and $a_2 = b_2+c_2$ being similar representations of two elements a_1, a_2 of A we have $a_1 a_2 = b_1 b_2 + (b_1 c_2 + b_2 c_1 + c_1 c_2)$. Hence $d(a_1 a_2) = h(b_1 c_2 + b_2 c_1 + c_1 c_2) \stackrel{(6)}{=}$
$\stackrel{(6)}{=} h(b_1 c_2 + b_2 c_1) = b_1 h(c_2) + b_2 h(c_1) \stackrel{(7)}{=} b_1 h(c_2) + c_1 h(c_2) + b_2 h(c_1) + c_2 h(c_1) =$
$= (b_1+c_1)h(c_2) + (b_2+c_2)h(c_1) = a_1 h(c_2) + a_2 h(c_1) = a_1 d(a_2) + a_2 d(a_1)$,
where (6) holds since $c_1 c_2 \in \mathfrak{m}^2 = (0)$ and (7) holds since $c_1 h(c_2) = 0 = c_2 h(c_1)$ since $c_1, c_2 \in \mathfrak{m}$. Thus $d \in \text{Der}_k(A, k)$. Evidently $\delta'(d) = d|\mathfrak{m} = h$. Hence δ' is surjective as required. This completes the proof of the proposition.

3.2. Proposition. Let k be a field and K be a finitely generated field extension of k. Then $\Gamma_k^1(K)$ is a K-vector space of dimension $\geq$ tr.deg. K/k and equality holds if and only if K is a separable field extension of k.

Proof.

Since $\Gamma_k^1(K) \simeq \operatorname{Hom}_K(\Gamma_k^1(K), K) \simeq \operatorname{Der}_k(K, K)$ as K-vector spaces by Prop. 2.4 and since $\operatorname{Der}_k(K, K)$ is a K-vector space of dimension $\geq$ tr.deg. K/k and equality holds if and only if K is a separable field extension of k (See Thm. 41, p. 127 in Vol. 1 of <u>Commutative Algebra</u> by Zariski & Samuel for a proof of the latter.) we obtain the conclusion of the proposition.

3.3. Definition.

Let k be a field, B be a k-algebra, $Q \in \operatorname{Spec}(B)$ and put $A = B_Q$. Put $n = \sup_P$ tr.deg. $\kappa(P)/k$ where P runs through the set of minimal prime ideals in A. We say <u>A is simple over k</u> or <u>B is simple over k at Q</u> if and only if there exists $f \in B$ such that $f \notin Q$ (or equivalently, such that Q belongs to the open subset $D(f)$ of $\operatorname{Spec}(B)$) and such that B_f is a finitely generated k-algebra and such that the minimum cardinality of a set of generators for $\Gamma_k^1(A)$ as an A-module is $\leq n$. We say <u>B is simple over k</u> if and only if B is simple over k at Q for each $Q \in \operatorname{Spec}(B)$.

3.3.1. Remark. With notation and hypotheses as in Def. 3.3 if B is simple over k, it follows that B is a finitely generated k-algebra.

3.3.2. Remark. See (2) and (3) in Remark 1.3.1 regarding alternate characterizations of the integer n in Def. 3.3.

Before proceeding to Theorem 3.5 we first need to prove Lemma 3.4 which is Theorem 3.5 under the additional hypothesis that the residue class field of A is a separable field extension of k.

3.4. Lemma. Let k be a field, B be a k-algebra, $Q \in \text{Spec}(B)$, put $A = B_Q$, let n be defined as in Def. 3.3 and let K denote the residue class field of A. Suppose A is simple over k and K is a separable field extension of k. Then A is a regular local ring (and therefore an integral domain), the inequality in Def. 3.3 is an equality and the quotient field L of A is a separable field extension of k.

Proof.

Let $\mathfrak{m}$ denote the maximal ideal of A and put $r = \dim A$. Choose $P \in \text{Spec}(A)$ such that $\dim A = \dim A/P$ (Such a prime ideal P exists by Thm. A.6 in the appendix.). Again by (4) in Thm. A.6 (in the appendix) we have

(1) $\qquad \text{tr.deg. } \kappa(P)/k = n$.

In order to show A is a regular local ring we need to show $\dim_K \mathfrak{m}/\mathfrak{m}^2 = r$.

Let $\mathfrak{n}$ denote the maximal ideal of A/P. But
(2) $\mathfrak{m}/\mathfrak{m}^2 = \mathfrak{n}/\mathfrak{n}^2$ and K is also the residue class field of A/P

and the homomorphisms of rings $k \longrightarrow A \longrightarrow A/P$ yield by Prop. 2.10 that

(3) $\begin{cases} P/P^2 \longrightarrow \Gamma^1_k(A) \underset{A}{\otimes} (A/P) \longrightarrow \Gamma^1_k(A/P) \longrightarrow 0 \\ \\ \text{is an exact sequence of A/P-modules.} \end{cases}$

In view of the inequality in Def. 3.3, (1) and (3) we conclude that

(4) $\left\{\begin{array}{l} \text{The minimum cardinality of a set of generators} \\ \text{for } \Gamma_k^1(A/P) \text{ as an } A/P\text{-module is} \\ \leq \text{tr.deg. } \kappa(P)/k. \end{array}\right.$

In view of (2) and (4), to show $\dim_K \mathfrak{m}/\mathfrak{m}^2 = r$ and complete the proof that A is a regular local ring we may assume that A is an integral domain by replacing A by A/P. Moreover, since K is also the residue class field of A/P and $\kappa(P)$ is the quotient field of A/P, to establish the last two conclusions of the lemma we may assume A is an integral domain by replacing A by A/P.

Put $m = \text{tr.deg. } K/k$ and let n_0 be the minimum cardinality of a set of generators for $\Gamma_k^1(A)$ as an A-module. Note that $n = \text{tr.deg. } L/k$. By (3) in Thm. A.6 (in the appendix) we have

(5) $\qquad\qquad\qquad r+m = n$.

Since A contains the field k and K is a separable field extension of k, by Prop. 3.1 we conclude

(6) $\left\{\begin{array}{l} 0 \longrightarrow \mathfrak{m}/\mathfrak{m}^2 \longrightarrow \Gamma_k^1(A) \underset{A}{\otimes} K \longrightarrow \Gamma_k^1(K) \longrightarrow 0 \\ \text{is an exact sequence of } K\text{-vector spaces.} \end{array}\right.$

In view of (6) we conclude

(7) $$\begin{cases} \dim_K \mathfrak{m}/\mathfrak{m}^2 = n_0 - \dim_K \Gamma_k^1(K) \overset{(8)}{\leq} n_0 - m = \\ n_0 - (n-r) = (n_0-n) + r \overset{(9)}{\geq} r \, , \end{cases}$$

where (8) holds since $\dim_K \Gamma_k^1(K) \geq m$ by Prop. 3.2 and where (9) holds since $n_0 \leq n$ by the inequality in Def. 3.3. Thus

(10) $$\dim_K \mathfrak{m}/\mathfrak{m}^2 \leq r \, .$$

But by Cor. A.3.1 (in the appendix) we obtain $\dim_K \mathfrak{m}/\mathfrak{m}^2 \geq r$. In view of (10) we conclude A is a regular local ring. Hence in view of (7) we have $r = \dim_K \mathfrak{m}/\mathfrak{m}^2 = (n_0-n) + r$ whence

(11) $$n_0 = n \, .$$

Hence the inequality in Def. 3.3 is an equality.

Since $\Gamma_k^1(L) = \Gamma_k^1(A)_{(0)}$ and since the inequality in Def. 3.3 is an equality we conclude $\dim_K \Gamma_k^1(L) \leq n = \text{tr.deg. } L/k$ whence by Prop. 3.2 we conclude that L is a separable field extension of k. This establishes the lemma.

3.5. Theorem. With notation and hypotheses as in Lemma 3.4, except we do not assume K is a separable field extension of k, the conclusions of Lemma 3.4 are true.

Proof.

If $ch(k) = 0$, K is automatically a separable field extension of k. Hence we may assume that $ch(k) = p > 0$. Put $m = \text{tr.deg. } K/k$, $k' = k^{p^{-\infty}}$

and let n_0 be the minimum cardinality of a set of generators for $\Gamma_k^1(A)$ as an A-module. Since A is simple over k there exists $f \in B$ such that $f \notin Q$ and such that B_f is a finitely generated k-algebra. By replacing B by B_f, we may assume without loss of generality that B is a finitely generated k-algebra. Hence there exists $N \geq 1$ and an ideal I in $k[T_1, \ldots, T_N]$ such that $B = k[T_1, \ldots, T_N]/I$. Put $I' = I \otimes_k k'$, i.e. I' is the ideal generated by the image of I in $k'[T_1, \ldots, T_N]$. Put $B' = k'[T_1, \ldots, T_N]/I'$. Note $B' = B \otimes_k k'$. Since k and k' are fields, the natural homomorphism $B \to B'$ is injective. Since B' is also integral over B there is a prime ideal Q' in B' lying over Q. Thus we obtain a local homomorphism of local rings $A = B_Q \to B'_{Q'}$. Put $A' = B'_{Q'}$ and let $\mathfrak{m}'$ denote the maximal ideal of the local ring A'.

Let $P' \in \text{Spec}(A')$ and put $P = P' \cap A$. Then we have a local homomorphism of local rings $A_P \to A'_{P'}$, which induces an injection on the residue class fields $\kappa(P) \to \kappa(P')$. Hence we have the commutative diagram (1) of field extensions.
Since k' is algebraic over k we conclude tr.deg. $\kappa(P')/k' \geq$
$\geq$ tr.deg. $\kappa(P)/k$. Hence putting

$$\begin{array}{ccc} k' & \longrightarrow & \kappa(P') \\ \uparrow & (1) & \uparrow \\ k & \longrightarrow & \kappa(P) \end{array}$$

$n' = \sup_{P'} \text{tr.deg. } \kappa(P')/k$, where
P' runs through the set of minimal prime ideals P' in A' we conclude $n' \geq n$. Let n'_0 be the minimal cardinality of a set of generators for $\Gamma_{k'}^1(A')$ as an A'-module. Note

(2) $$\begin{cases} \Gamma^1_{k'}(A') = \Gamma^1_{k'}(B'_{Q'}) = \Gamma^1_{k'}(B') \underset{B'}{\otimes} B'_{Q'} = \\ = \Gamma^1_{k'}(B \underset{k}{\otimes} k') \underset{B'}{\otimes} B'_{Q'} = (\Gamma^1_k(B) \underset{k}{\otimes} k') \underset{B'}{\otimes} B'_{Q'} = \\ = (\Gamma^1_k(B_Q) \underset{k}{\otimes} k') \underset{B'}{\otimes} B'_{Q'} = (\Gamma^1_k(A) \underset{k}{\otimes} k') \underset{B'}{\otimes} A' \end{cases}$$

and

(3) $$\begin{cases} (A^{n_0} \underset{k}{\otimes} k') \underset{B'}{\otimes} A' = (A \underset{k}{\otimes} k')^{n_0} \underset{B'}{\otimes} A' = \\ = ((A \underset{k}{\otimes} k') \underset{B'}{\otimes} A')^{n_0} = ((B \underset{k}{\otimes} k')_Q \underset{B'}{\otimes} B'_{Q'})^{n_0} = \\ = (B'_Q \underset{B'}{\otimes} B'_{Q'})^{n_0} = B'^{n_0}_{Q'} = A'^{n_0} . \end{cases}$$

By definition of n_0 we have an epimorphism of A-modules $A^{n_0} \longrightarrow \Gamma^1_k(A)$. Tensoring this epimorphism over k with k' yields an epimorphism $A^{n_0} \underset{k}{\otimes} k' \longrightarrow \Gamma^1_k(A) \underset{k}{\otimes} k'$ and tensoring the latter epimorphism over B' with A' yields an epimorphism $(A^{n_0} \underset{k}{\otimes} k') \underset{B'}{\otimes} A' \longrightarrow (\Gamma^1_k(A) \underset{k}{\otimes} k') \underset{B'}{\otimes} A'$ of A'-modules and in view of (2) and (3) this is an epimorphism $A'^{n_0} \longrightarrow \Gamma^1_{k'}(A')$ of A'-modules. Hence $n'_0 \le n_0$. Thus $n'_0 \le n_0 \le n \le n'$. The residue class field of A' is a separable field extension of k' since the field k' is perfect. Hence by Lemma 3.4 we conclude that A' is a regular local ring.

Since k is a field, $B \otimes k'$ is a flat B-module whence $B_Q \underset{k}{\otimes} k'$ is a flat B_Q-module. But $(B_Q \underset{k}{\otimes} k') \underset{B'}{\otimes} B'_{Q'}$ is a flat $B_Q \underset{k}{\otimes} k'$-module. Hence $(B_Q \underset{k}{\otimes} k') \underset{B'}{\otimes} B'_{Q'}$ is a flat B_Q-module. But $A = B_Q$ and $A' = B'_{Q'} = (B \underset{k}{\otimes} k')_{Q'} = (B_Q \underset{k}{\otimes} k') \underset{B'}{\otimes} B'_{Q'}$. Thus we have shown A' is a flat A-module. Hence by Lemma A.4 (in the appendix) we conclude A is a regular local ring.

We have $n_0 \geq \dim_L \Gamma_k^1(L) \geq \text{tr.deg. } L/k = n$, where the first inequality holds by the inequality in Def. 3.3 and the second inequality holds by Prop. 3.2. Since also $n_0 \leq n$ we conclude $n_0 = n$, that is the inequality in Def. 3.3 is an equality and that $\dim_L \Gamma_k^1(L) = \text{tr.deg. } L/k$ which by Prop. 3.2 implies that L is a separable field extension of k. This completes the proof of the theorem.

CHAPTER 4

Brief discussion of unramified and étale homomorphisms

In this chapter we present only the definitions and a few very basic properties of unramified and étale homomorphisms needed in the succeeding chapters in the treatment of simple homomorphisms. Those wishing to learn more of the basic properties of unramified and étale homomorphisms should consult one of the many current texts in algebraic geometry.

4.1. Proposition. Let k be a field, B be a k-algebra, $Q \in \mathrm{Spec}(B)$ and put $A = B_Q$. Suppose there exists $f \in B$ such that $f \notin Q$ and such that B_f is a finitely generated k-algebra and suppose $A \neq 0$. Then the following two conditions are equivalent:

(1) $\Gamma_k^1(A) = 0$.

(2) A is a field and A is a finite separable field extension of k.

Proof.

(1) $\Longrightarrow$ (2). Let K denote the residue class field of A. By replacing B by B_f we may assume without loss of generality that B is a finitely generated k-algebra. By Prop. 2.10 we have an exact sequence of homomorphisms of K-vector spaces $\Gamma_k^1(A) \otimes_A K \longrightarrow \Gamma_k^1(K) \longrightarrow 0$ whence $\Gamma_k^1(K) = 0$ since $\Gamma_k^1(A) \otimes_A K = 0$ by (1). Hence by Prop. 3.2 we conclude that K is a finite separable algebraic extension of k.

Since $\Gamma_k^1(K) = 0$, by Lemma 3.4 we conclude that A is a regular local ring and, in particular, an integral domain. Let L denote the quotient field of A. By (1) we obtain $\Gamma_k^1(L) = 0$ whence by Prop. 3.2 we conclude that L is an algebraic extension of k. Hence by (3) in Thm. A.6 (in the appendix) we conclude that dim A = 0 whence A is a field. This establishes (2).

(2) $\Longrightarrow$ (1). (2) and Prop. 3.2 yield (1).

4.2.1. Definition. Let k be a field, B be a k-algebra, $Q \in \text{Spec}(B)$ and put $A = B_Q$. We say that <u>A is étale over k</u> or <u>B is étale over k at Q</u> if and only if there exists $f \in B$ such that $f \notin Q$ and B_f is a finitely generated k-algebra and if condition (1) in Prop. 4.1 is satisfied. We say that <u>B is étale over k</u> if and only if B is étale over k at Q for all $Q \in \text{Spec}(B)$.

4.2.2. Definition. Let A be a ring, B be an A-algebra, $\lambda : A \longrightarrow B$ be the structure homomorphism, $Q \in \text{Spec}(B)$, $P = \lambda^{-1}(Q)$ and let Q' denote the image of Q in $B \otimes_A \kappa(P) = B/m_{A_P} B$. We say that <u>B is unramified over A at Q</u> if and only if $B \otimes_A \kappa(P)$ is étale over $\kappa(P)$ at Q' and there exists $f \in B$ such that $f \notin Q$ and B_f is a finitely presented A-algebra. We say that <u>B is étale over A at Q</u> if and only if B is unramified over A at Q and B_Q is a flat A_P-module. We say <u>B is unramified over A</u> or <u>λ is unramified</u> (resp., <u>B is étale over A</u> or <u>λ is étale</u>) if and only if B is unramified over A at Q (resp., B is étale over A at Q) for all $Q \in \text{Spec}(B)$.

4.2.3. Remark. The condition "there exists $f \in B$ such that $f \notin Q$ and B_f is a finitely presented A-algebra" in Def. 4.2.2 is not superfluous. For example, take $A = \mathcal{O}$ and $B = K$ where $\mathcal{O}$ is a discrete valuation ring and K is the quotient field of $\mathcal{O}$. Then $K \otimes_{\mathcal{O}} K = K$, $K \otimes_{\mathcal{O}} k = 0$ where k is the residue class field of $\mathcal{O}$, K is étale over K and 0 is étale over k. But K is a finitely presented $\mathcal{O}$-algebra if and only if $\mathcal{O}$ is a field.

4.2.4. Remark. Let A be a ring and B be an A-algebra. If B is unramified over A then B is a finitely presented A-algebra; if B is étale over A then B is a flat finitely presented A-algebra.

4.3. Proposition. Let A be a ring, B be an A-algebra with structure homomorphism $\lambda : A \longrightarrow B$, let $Q \in \mathrm{Spec}(B)$, put $P = \lambda^{-1}(Q)$, $A' = A_P$ and $B' = B_Q$. Let K (resp., L) denote the residue class field of A' (resp., B'). Suppose there exists $f \in B$ such that $f \notin Q$ and such that B_f is a finitely presented A-algebra. Then the following conditions are equivalent:

(1A) $\Gamma^1_{A'}(B') = 0$.
(1B) $\Gamma^1_A(B') = 0$.
(1C) $\Gamma^1_A(B)_Q = 0$.
(2) $\mathfrak{m}_{A'}B' = \mathfrak{m}_{B'}$ and L is a finite separable field extension of K.
(3) B is unramified over A at Q.

Proof.

(1A) $\Longleftrightarrow$ (1B) $\Longleftrightarrow$ (1C). Immediate by Cor. 2.13.1.

Let Q' be the image of Q in $C = B_P/PB_P$ and put $C' = C_{Q'} = B'/\mathcal{m}_{A'}B'$. Note that C_f is a finitely generated K-algebra and $\Gamma^1_K(C') = \Gamma^1_K(B' \underset{A'}{\otimes} K) = \Gamma^1_{A'}(B') \underset{A'}{\otimes} K = \mathcal{m}_{A'}\Gamma^1_{A'}(B') = \mathcal{m}_{A'}B'\Gamma^1_{A'}(B')$.

(1A) $\Longrightarrow$ (2). By (1A) we obtain $\Gamma^1_K(C') = \Gamma^1_{A'}(B') \underset{A'}{\otimes} K = 0$. Hence by Prop. 4.1 we obtain (2).

(2) $\Longrightarrow$ (3). By definition, to establish (3) we need to show $C_Q = C'$ is étale over K. But $C' = B'/\mathcal{m}_{A'}B' = B'/\mathcal{m}_{B'} = L$ since $\mathcal{m}_{A'}B' = \mathcal{m}_{B'}$ by (2). Thus C' is a field and by (2) we have C' is a finite separable field extension of K. Hence by Prop. 4.1 we conclude that C' is étale over K, which establishes (3).

(3) $\Longrightarrow$ (1A). By (3) and by definition we have C' is étale over K. Hence by definition we have $\Gamma^1_K(C') = 0$. By Prop. 4.1 we conclude that C' is a field. Hence $\mathcal{m}_{A'}B' = \mathcal{m}_{B'}$. Hence $\mathcal{m}_{B'}\Gamma^1_{A'}(B') = \mathcal{m}_{A'}B'\Gamma^1_{A'}(B') = \Gamma^1_K(C') = 0$. Since $\Gamma^1_{A'}(B') = \Gamma^1_A(B)_Q$ is a finitely generated $B_Q = B'$-module, by Nakayama's lemma we obtain (1A).

4.4. Proposition. Let A be a ring and B be an A-algebra. Then the set of points in $\text{Spec}(B)$ at which B is unramified over A (resp., étale over A) is an open subset of $\text{Spec}(B)$ (with respect to the Zariski topology).

Proof.

Suppose B is unramified over A at Q for some $Q \in \text{Spec}(B)$. Hence by Prop. 4.3 we conclude $\Gamma^1_A(B)_Q = 0$. Since B is unramified over A at Q there exists $g \in B$ such that $g \notin Q$ and such that

B_g is a finitely presented A-algebra. Hence $\Gamma_A^1(B_g)$ is a finitely presented B_g-module by Cor. 2.10.2. Since $\Gamma_A^1(B_g)$ is a finitely presented B_g-module, by Lemma A.14 (in the appendix) we conclude that there exists $f \in B$ such that $f \notin Q$ and such that $\Gamma_A^1(B)_f = 0$. Hence given $P \in \operatorname{Spec}(B)$ such that $P \in D(f)$ (i.e. such that $f \notin P$) we have $\Gamma_A^1(B)_P = (\Gamma_A^1(B)_f)_P = 0$ whence by Prop. 4.3 we conclude B is unramified over A at P.

On the other hand, suppose for some $Q \in \operatorname{Spec}(B)$, letting P denote the preimage of Q in A under the structure homomorphism, that B_Q is a flat A_P-module and that there exists $g \in B$ such that $g \notin Q$ and such that B_g is a finitely presented A-algebra. Then by Thm. A.7 (in the appendix) we conclude that there exists $f \in B$ such that $Q \in D(f)$ and such that for each $Q_1 \in D(f)$, letting P_1 denote the preimage of Q_1 in A under the structure homomorphism, B_{Q_1} is a flat A_{P_1}-module. This completes the proof of the proposition.

4.5. Lemma. Let k be a field and K be a finitely generated field extension of k. Then the following two conditions are equivalent:

(1) K is a separable field extension of k.

(2) There exists an integer $n \geq 0$ and an étale
homomorphism of k-algebras $\lambda : k[T_1, \ldots, T_n] \longrightarrow K$.

(Note that the proof shows in the case of (2) that we necessarily have $n = \operatorname{tr.deg.} K/k$ and $\lambda^{-1}(0) = 0$ for any such homomorphism λ whence $k(\lambda(T_1), \ldots, \lambda(T_n))$ is purely transcendental over k.)

Proof.

(1) $\implies$ (2). Since K is a finitely generated field extension of k, in view of (1) there exists a separating transcendence basis $\{x_1, \ldots, x_n\}$ for K over k. Put $L = k(x_1, \ldots, x_n)$. Therefore K is a separable algebraic field extension of L and therefore K is also necessarily a finite field extension of L since K is a finitely generated field extension of k. Define a homomorphism $\lambda : k[T_1, \ldots, T_n] \longrightarrow K$ of k-algebras by $\lambda(T_i) = x_i$ for $1 \le i \le n$. Then $\lambda^{-1}(0) = 0$ since L is transcendental over k. Since K is a flat L-module and since K is a finite separable field extension of L, by Prop. 4.3 we conclude that λ is an étale homomorphism. This establishes (2).

(2) $\implies$ (1). Let n and λ be as in (2). Put $P = \lambda^{-1}(0)$ and $A = k[T_1, \ldots, T_n]$. Since K is an étale A-algebra we conclude by definition that K is a flat A_P-module. Hence by Prop. A.8 (in the appendix) with $P_1 = (0)$ and $Q = (0)$ we conclude $P = (0)$. Hence $\kappa(P) = k(T_1, \ldots, T_n)$. Since K is an étale A-algebra, by Prop. 4.3 we conclude K is a finite separable field extension of $\kappa(P)$. Hence tr.deg. K/k = tr.deg $\kappa(P)/k = n$. $\kappa(P)$ is a separable field extension of k, being a purely transcendental extension. Hence K is a separable field extension of k. This establishes the lemma.

4.6. Proposition. Let A be a ring, B be an A-algebra with structure homomorphism $\alpha : A \longrightarrow B$, C be a B-algebra with structure homomorphism $\beta : B \longrightarrow C$, let $R \in \operatorname{Spec}(C)$, put $Q = \beta^{-1}(R)$ and $P = \alpha^{-1}(Q)$. Suppose C is unramified (resp., étale) over B at R and B is unramified (resp., étale) over A at Q. Then C is unramified (resp., étale) over A at R.

Proof.

In the unramified case, by Prop. 4.3 the hypotheses mean

(1) $\quad\quad\quad \Gamma^1_{B_Q}(C_R) = 0 \quad \text{and} \quad \Gamma^1_A(B_Q) = 0$.

In view of Prop. 2.9 we have an exact sequence of C_R-modules

(2) $\quad\quad \Gamma^1_A(B_Q) \underset{B_Q}{\otimes} C_R \longrightarrow \Gamma^1_A(C_R) \longrightarrow \Gamma^1_{B_Q}(C_R) \longrightarrow 0$.

In view of (1) and (2) we obtain $\Gamma^1_A(C_R) = 0$ which by Prop. 4.3 yields that C is unramified over A at R.

On the other hand, supposing that C_R is a flat B_Q-module and that B_Q is a flat A_P-module it remains only to show that C_R is a flat A_P-module, but this is well-known.

4.6.1. Corollary. Let A be a ring, B be an A-algebra and C be a B-algebra. Suppose C is unramified (resp., étale) over B and that B is unramified (resp., étale) over A. Then C is unramified (resp., étale) over A.

Proof.

Immediate by definition and by Prop. 4.6.

4.7. Proposition. Let A be a ring, B and C be A-algebras, let $\beta : A \longrightarrow C$ be the structure homomorphism, put

$\beta' = \beta \otimes_A B : B \longrightarrow B \otimes_A C$, let $R \in \mathrm{Spec}(B \otimes_A C)$ and put $Q = \beta'^{-1}(R)$. Suppose B is unramified (resp., étale) over A at Q. Then $B \otimes_A C$ is unramified (resp., étale) over C at R.

Proof.

In the unramified case, by Prop. 4.3 we have $\Gamma^1_A(B_Q) = 0$. Hence
$$\Gamma^1_C((B \otimes_A C)_R) = \Gamma^1_C((B_Q \otimes_A C)_R) = (\Gamma^1_C(B_Q \otimes_A C))_R = (\Gamma^1_A(B_Q) \otimes_A C)_R = 0$$
whence by Prop. 4.3 we conclude $B \otimes_A C$ is unramified over C at R.

On the other hand, suppose B_Q is a flat A_P-module, where $\alpha : A \longrightarrow B$ is the structure homomorphism and $P = \alpha^{-1}(Q)$. Put $\alpha' = \alpha \otimes_A C : C \longrightarrow B \otimes_A C$ and $R_0 = \alpha'^{-1}(R)$. $(B \otimes_A C)_R = (B_Q \otimes_{A_P} C_{R_0})_R$ is evidently a flat $B_Q \otimes_{A_P} C_{R_0}$-module, and since B_Q is a flat A_P-module we conclude $B_Q \otimes_{A_P} C_{R_0}$ is a flat C_{R_0}-module whence $(B \otimes_A C)_R$ is a flat C_{R_0}-module. This completes the proof of the proposition.

4.7.1. Corollary. Let A be a ring and B and C be A-algebras. Suppose B is unramified (resp., étale) over A. Then $B \otimes_A C$ is unramified (resp., étale) over C.

Proof.

Immediate by Prop. 4.7.

4.8. Proposition. Let A be a ring, B be an A-algebra and C be a B-algebra. Suppose C is unramified (resp., étale) over B. Then the canonical homomorphism $\lambda : \Gamma^1_A(B) \otimes_B C \longrightarrow \Gamma^1_A(C)$ from the exact sequence in Prop. 2.9 is surjective (resp., bijective).

Proof.

If C is unramified over B then by Prop. 4.3 we conclude that $\Gamma^1_B(C)_Q = 0$ for all $Q \in \text{Spec}(C)$ whence $\Gamma^1_B(C) = 0$. Hence by Prop. 2.9 we conclude that the canonical homomorphism λ is surjective.

Now assuming, in addition, that C is a flat B-module, it suffices in view of the above paragraph to prove that λ is injective. With notation as in Def. 2.1 put $I = I_{C/A}$, let $\gamma : B \otimes_A C \longrightarrow C$ be the homomorphism defined by $\gamma(b \otimes c) = bc$ for all $b \in B$ and $c \in C$ and put $G = \text{Ker}(\gamma)$. The natural homomorphism $B \otimes_A C \longrightarrow C \otimes_A C$ is injective since C is a flat B-module. Hence $G \subset I$. Let $\gamma' : B \otimes_A C/G^2 \longrightarrow C$ (resp., $\pi'_{B/A} : B \otimes_A B/I^2_{B/A} \longrightarrow B$) be the homomorphism obtained from γ (resp., $\pi_{B/A}$). Then $\Gamma^1_A(B) \otimes_B C = \text{Ker}(\pi'_{B/A}) \otimes_B C = \text{Ker}(\gamma')$ since C is a flat B-module. Hence λ is the natural homomorphism $G/G^2 \longrightarrow I/I^2$. Let $\varepsilon : C \otimes_A C \longrightarrow C \otimes_B C$ be the natural homomorphism and put $H = \text{Ker}(\varepsilon)$. Since $G \subset H \cap I$ and since $H \cap I^2 \subset H^2$ (since $I^2_{C/B} = I_{C/B}$ since $\Gamma^1_B(C) = 0$) we obtain $G \cap I^2 \subset G^2$ which proves λ is injective. This establishes the proposition.

CHAPTER 5

Some corollaries to Theorem 3.5

This chapter consists of several corollaries to Theorem 3.5 in Chapter 3; they were postponed to this chapter since material in Chapter 4 is needed.

5.1. Corollary. With notation and hypotheses as in Thm. 3.5 we conclude that for all $P \in \text{Spec}(A)$ the following are true:

(1) The minimum cardinality for a set of generators for
$\Gamma_k^1(A_P)$ as an A_P-module $= \text{tr.deg. } L/k$.

(2) A_P is a regular local ring.

Proof.
Since $\Gamma_k^1(A_P) = \Gamma_k^1(A)_P$ we conclude by hypothesis that $\Gamma_k^1(A_P)$ can be generated as an A_P-module by a subset with $\leq \text{tr.deg. } L/k$ elements. Hence by Thm. 3.5 with A_P in place of A we obtain the conclusions of the corollary.

5.2. Corollary. With notation and hypotheses as in Thm. 3.5 we conclude that $\Gamma_k^1(A)$ is a free A-module of rank $\text{tr.deg. } L/k$.

Proof.
It is immediate by (1) of Cor. 5.1 that $\Gamma_k^1(A)$ is a locally free A-module of rank $\text{tr.deg. } L/k$ and therefore a free module of this rank since A is a local ring.

5.3. Corollary. With notation and hypotheses as in Thm. 3.5 we conclude that L is a separable field extension of k.

Proof.
This was observed in the proof of Thm. 3.5.

5.4. Corollary. With notation and hypotheses as in Thm. 3.5, putting n = tr.deg. L/k, there exists a subset $\{x_1, \ldots, x_n\}$ of A which is algebraically independent over k such that $\Gamma_k^1(A)$ is a free A-module with basis $\{d(x_1), \ldots, d(x_n)\}$.

Proof.

By the normalization lemma there exists a subset $\{x_1, \ldots, x_n\}$ of A such that A is integral over $k[x_1, \ldots, x_n]$. Thus L is algebraic over $k(x_1, \ldots, x_n)$. Since L is a separable field extension of k by Cor. 5.3 and since $\{x_1, \ldots, x_n\}$ is a transcendence basis for L over k, $\{x_1, \ldots, x_n\}$ is already a separating transcendence basis for L over k. By Lemma 4.5 and its proof we conclude that

(1) $$\Gamma_C^1(L) = 0 \;,$$

where $C = k[x_1, \ldots, x_n]$.

The homomorphisms of rings $k \longrightarrow C \longrightarrow L$ yield by Prop. 2.9 an exact sequence of homomorphisms of L-vector spaces

(2) $$\Gamma_k^1(C) \otimes_C L \longrightarrow \Gamma_k^1(L) \longrightarrow \Gamma_C^1(L) \longrightarrow 0 \;.$$

In view of (1) and (2) we have an epimorphism $\lambda : \Gamma_k^1(C) \otimes_C L \longrightarrow \Gamma_k^1(L)$ of L-vector spaces. But $\{d(x_1), \ldots, d(x_n)\}$ is a basis for the C-module $\Gamma_k^1(C)$ by Prop. 2.8 since $\{x_1, \ldots, x_n\}$ is algebraically independent over k. Hence $\{d(x_1) \otimes 1, \ldots, d(x_n) \otimes 1\}$ is a basis for the L-vector space $\Gamma_k^1(C) \otimes_C L$. Thus

(3) $\begin{cases} \{\lambda(d(x_1) \otimes 1), \ldots, \lambda(d(x_n) \otimes 1)\} = \{d(x_1), \ldots, d(x_n)\} \\ \text{is a set of generators for the L-vecor space } \Gamma_k^1(L). \end{cases}$

Since $\Gamma_k^1(A)$ is a free A-module of rank n by Cor. 5.2 and since $\Gamma_k^1(L) = \Gamma_k^1(A)_{(0)}$, we conclude

(4) $\quad \Gamma_k^1(L)$ is an L-vector space of dimension n.

In view of (3) and (4) we obtain

(5) $\begin{cases} \{d(x_1), \ldots, d(x_n)\} \text{ is a basis for the L-vector} \\ \text{space } \Gamma_k^1(L). \end{cases}$

But $\{d(x_1), \ldots, d(x_n)\} \subset \Gamma_k^1(A)$ since $\{x_1, \ldots, x_n\} \subset A$. Hence in view of (5) we conclude $\{d(x_1), \ldots, d(x_n)\}$ is a basis for the free A-module $\Gamma_k^1(A)$. This establishes the corollary.

CHAPTER 6

Fitting ideals

Let A be a ring, M be a finitely presented A-module and choose a presentation $(*) : A^m \longrightarrow A^n \longrightarrow M \longrightarrow 0$ for M as an A-module, $m, n \geq 0$ (i.e. $(*)$ is an exact sequence of A-modules). In Def. 6.1 below for each integer $r \geq 0$ we define the rth-fitting ideal $I_r^{A,(*)}(M)$ of the A-module M (with respect to the presentation $(*)$). Later in this chapter we are able to establish Thm. 6.6 and Cor. 6.6.1 which shall enable us to establish the Jacobian criterion in Thm. 7.1 in Chapter 7.

6.1. Definition. Let A be a ring and M be a finitely presented A-module. Since M is a finitely presented A-module there exists an exact sequence of homomorphisms of A-modules
$(*) : A^m \longrightarrow A^n \longrightarrow M \longrightarrow 0$ for some integers $m, n \geq 1$. Let $a_0 = (a_{j,i})_{1 \leq i \leq m, 1 \leq j \leq n}$ be the matrix associated to the homomorphism $A^m \longrightarrow A^n$ of free A-modules with respect to the standard bases and let a be the square matrix obtained from a_0 by adding either enough columns of zeros to the right or enough rows of zeros to the bottom, as the case requires.

For each integer $r \geq 0$ we define the rth-fitting ideal of the A-module M with respect to the presentation $(*)$, denoted

$I_r^{A,(*)}(M)$, or simply I_r when both A, (*) and M are understood, as the ideal in A generated by all (n-r) × (n-r) minors of the matrix a (which are $\binom{n}{n-r}\binom{m}{n-r}$ in number).

6.1.1. Remark. With hypotheses and notation as in Def. 6.1 we have
$I_{n-1} = \sum_{i=1}^{m}\sum_{j=1}^{n} a_{j,i} A$, $I_n = A$ and $I_0 \subset I_1 \subset \ldots \subset I_{n-1} \subset I_n = A$.

6.1.2. Remark. With hypotheses and notation as in Def. 6.1 we ignore the question of whether $I_r^{A,(*)}(M)$, $r \geq 0$ are, in general, independent of the presentation (*) for M since we do not need to prove independence of the presentation to establish Thm. 6.6 and Cor. 6.6.1, although doing so would simplify the exposition.

6.2. Proposition. Let A be a ring, M be a finitely presented A-module and let S be a multiplicatively closed subset of A. Fix a presentation (*) : $A^m \longrightarrow A^n \longrightarrow M \longrightarrow 0$, m, n $\geq$ 0 for M as an A-module. Let (**) : $(S^{-1}(A))^m \longrightarrow (S^{-1}(A))^n \longrightarrow S^{-1}(M) \longrightarrow 0$

be the presentation for $S^{-1}(M)$ as an $S^{-1}(A)$-module obtained from (*) by passing to quotients.

Then $I_r^{A,(*)}(M)\cdot(S^{-1}(A)) = I_r^{S^{-1}(A),(**)}(S^{-1}(M))$ for each integer $r \geq 0$.

Proof.

Let a_0 be defined as in Def. 6.1. Then a_0 is also the matrix associated to the homomorphism $(S^{-1}(A))^m \longrightarrow (S^{-1}(A))^n$ of free $S^{-1}(A)$-modules with respect to the standard bases. The conclusions of the proposition are now immediate by definition.

6.2.1. Corollary. Let A be a ring, M be a finitely presented A-module and let $P \in \text{Spec}(A)$. Fix a presentation (*) : $A^m \longrightarrow A^n \longrightarrow M \longrightarrow 0$, m, n ≥ 0 for M as an A-module and let (**) : $A_P^m \longrightarrow A_P^n \longrightarrow M_P \longrightarrow 0$ be the presentation for M_P as an A_P-module obtained from (*) by passing to quotients.

Then $I_r^{A,(*)}(M)_P = I_r^{A_P,(*)}(M_P)$ for each integer $r \geq 0$.

Proof.

Immediate by Prop. 6.2 by putting $S = A\backslash P$.

6.3. Lemma. Let A be a local ring, M be a finitely presented A-module and s be an integer ≥ 0. Suppose M can be generated as an A-module by a subset with s elements. Then there exists a presentation $A^m \longrightarrow A^s \longrightarrow M \longrightarrow 0$ for M as an A-module where m is an integer ≥ 0.

Proof.

Let $A^m \to A^n \xrightarrow{f} M \to 0$, $m, n \geq 0$ be a presentation for M as an A-module. Suppose first that $s > n$. For each $1 \leq i \leq s$ let $e_i \in A^s$ be the tuple with 1 in the ith-coordinate and all other coordinates 0. Define a homomorphism $g : A^s = A^n \times A^{s-n} \to M$ of A-modules by $g(a, b) = f(a)$ for $a \in A^n$ and $b \in A^{s-n}$. g is evidently surjective since f is surjective. Let $\{x_1, \ldots, x_{m_0}\} \subset A^n$ be a set of generators for $\text{Ker}(f)$ as an A-module. Then $\{(x_1, 0), \ldots, (x_{m_0}, 0), e_{n+1}, \ldots, e_s\}$ is a set of generators for $\text{Ker}(g)$ as an A-module. Since we can thus take $m = m_0 + s-n$ this establishes the lemma for the case $s > n$.

On the other hand, suppose $s \leq n$. For $1 \leq i \leq n$ let $e_i \in A^n$ be the tuple with 1 in the ith-coordinate and all other coordinates 0 and put $x_i = f(e_i)$. Since f is surjective $\{x_1, \ldots, x_n\}$ generates M as an A-module. By hypothesis there exists $\{y_1, \ldots, y_s\} \subset M$ such that M can be generated as an A-module by $\{y_1, \ldots, y_s\}$. Since A is a local ring, by Lemma A.20 (in the appendix) the tuple $y = (y_1, \ldots, y_s, 0, \ldots, 0)$ in M^n can be obtained from the tuple $x = (x_1, \ldots, x_n)$ in M^n by operations of the following types, letting $z = (z_1, \ldots, z_n)$ denote a tuple in M^n obtained from the tuple x by operations of the following types:

Type 1. For some $1 \leq i, j \leq n$ interchanging z_i and z_j.

Type 2. Replacing z_1 by az_1 for some unit $a \in A$.

Type 3. If $n \geq 2$, replacing z_1 by $z_1 + az_2$ for some $a \in A$.

Letting z denote a tuple in M^n as above, note that $Az_1 + \ldots + Az_n = M$ in view of the three types of operations, and define an epimorphism of A-modules $f_z : A^n \longrightarrow M$ by $f_z(e_i) = z_i$ for $1 \leq i \leq n$. Let $z, z' \in M^n$ be two elements of M^n as above and suppose z' is obtained from z by an operation of one of the three types. In case z' is obtained from z by an operation of Type 1 or Type 2 it is evident that if $\text{Ker}(f_z)$ is generated as an A-module by t elements, $t \geq 0$ then $\text{Ker}(f_{z'})$ is generated as an A-module by t elements, and conversely. Now suppose $n \geq 2$ and z' is obtained from z by an operation of Type 3. Hence $z_1' = z_1 + az_2$ for some $a \in A$ and $z_i' = z_i$ for $2 \leq i \leq n$. Let $t \geq 0$ and $\{a_1, \ldots, a_t\} \subset A^n$. Note that given $b = (b_1, \ldots, b_n) \in A^n$ that $b \in \text{Ker}(f_z)$ if and only if $(b_1-ab_2, b_2, \ldots, b_n) \in \text{Ker}(f_{z'})$. For each $1 \leq i \leq t$ put $a_i = (a_{i,1}, \ldots, a_{i,n})$ and put $a_i' = (a_{i,1}-aa_{i,2}, a_{i,2}, \ldots, a_{i,n})$. Hence $\{a_1, \ldots, a_t\}$ generates $\text{Ker}(f_z)$ as an A-module if and only if $\{a_1', \ldots, a_t'\}$ generates $\text{Ker}(f_{z'})$ as an A-module.

Since we have the exact sequence $A^{m_0} \longrightarrow A^n \xrightarrow{f} M \longrightarrow 0$ of homomorphisms of A-modules there exists a subset $\{a_1, \ldots, a_{m_0}\}$ of A^n which generates $\text{Ker}(f)$ as an A-module. Hence in view of the above observations we conclude that there exists a subset $\{b_1', \ldots, b_{m_0}'\}$ of A^n which generates $\text{Ker}(f_y)$ as an A-module.

Let $\pi : A^s \times A^{n-s} = A^n \longrightarrow A^s$ be the projection homomorphism defined by $\pi(a, b) = a$ for all $(a, b) \in A^s \times A^{n-s}$. For each $1 \leq i \leq m_0$ put $b_i = \pi(b_i')$. Let $\iota : A^s \longrightarrow A^n = A^s \times A^{n-s}$ be the homomorphism defined by $\iota(a) = (a, 0)$ for all $a \in A^s$. Then $\{b_1, \ldots, b_{m_0}\} \subset A^s$ generates $\text{Ker}(f_y \circ \iota)$. Since we can thus take $m = m_0$ this establishes the lemma for the case $s \leq n$ and hence completes the proof.

6.3.1. Corollary. Let A be a local ring, M be a finitely presented A-module and s be an integer ≥ 0. Suppose M can be generated as an A-module by a subset with s elements. Let
(*) : $A^m \longrightarrow A^n \longrightarrow M \longrightarrow 0$, $m, n \geq 0$ be a presentation for M as an A-module such that $n > s$.

Then there exists a presentation (**) : $A^m \longrightarrow A^s \longrightarrow M \longrightarrow 0$ for M as an A-module such that $I_r^{A,(*)}(M) = I_r^{A,(**)}(M)$ for each integer $r \geq 0$.

Proof.

By the proof of Lemma 6.3, putting $m = m_0$, with notation as in that proof we obtain a presentation (***) : $A^m \xrightarrow{g'} A^n \xrightarrow{f_y} M \longrightarrow 0$ for M as an A-module where the homomorphism g' is given by the matrix $\mathcal{b}' = (b'_{j,i})_{1 \leq j \leq n, 1 \leq i \leq m}$, where for all $1 \leq i \leq m$ we put $b'_i = (b'_{1,i}, \ldots, b'_{n,i})$ and we obtain a presentation
(**) : $A^m \xrightarrow{g} A^s \xrightarrow{f_y \circ \iota} M \longrightarrow 0$ for M as an A-module where the homomorphism g is given by the matrix $\mathcal{b} = (b'_{j,i})_{1 \leq j \leq s, 1 \leq i \leq m}$ and where

(1) $\begin{cases} \text{For each } s+1 \leq j \leq n, \text{ the } j^{\text{th}}\text{-row in the matrix} \\ \mathcal{b}' \text{ generates the unit ideal in } A. \end{cases}$

In view of (1) we can obtain a new matrix $b'' = (b''_{j,i})_{1 \leq j \leq n, 1 \leq i \leq m}$ from b' such that for each $s+1 \leq j \leq n$, $(b''_{j,1}, \ldots, b''_{j,m}) = (1, 0, \ldots, 0)$ by operations of the following types:

Type 1. For some unit $a \in A$ multiplying each member of a column of the matrix by a.

Type 2. For some $a \in A$ replacing a given column of the matrix by the sum of the given column and another column multiplied by a.

Since these operations do not alter the correponding fitting ideals, in order to show that the fitting ideals corresponding to b and b' coincide we may assume without loss of generality that $b' = b''$. But it is obvious now that the fitting ideals corresponding to b and b' coincide.

Let a be the matrix corresponding to the homomorphism $A^m \rightarrow A^n$ in the presentation (*). By the proof of Lemma 6.3 we conclude that the matrix b' can be obtained from the matrix a by operations of the following types:

Type 1. For some $1 \leq i, j \leq n$ interchanging the ith and j^{th} columns of the matrix.

Type 2. Replacing the first column of the matrix by some unit multiple of it.

Type 3. For some $a \in A$ replacing the first column of the matrix by the sum of the first column and another column multiplied by a.

Since these operations do not alter the corresponding fitting ideals we conclude that the fitting ideals corresponding to a and b' coincide. Thus the fitting ideals corresponding to a and b coincide. This is the conclusion of the corollary.

6.4. Lemma. Let A be a ring, r be an integer ≥ 0 and M be a free A-module of rank r. Then every epimorphism $A^r \longrightarrow M$ of A-modules is an isomorphism; in other words, every subset of M of cardinality r which generates M as an A-module is a basis for M as a free A-module.

Proof.

It suffices to consider only the case $M = A^r$. Let $\{a_1, \ldots, a_r\}$ be a subset of M which generates M as an A-module. For each $1 \leq i \leq r$ put $a_i = (a_{1,i}, \ldots, a_{r,i})$. Put $a = (a_{i,j})_{1 \leq i \leq r, 1 \leq j \leq r}$. Let I be the $r \times r$ identity matrix. Since $\{a_1, \ldots, a_r\}$ generates M as an A-module there exists a matrix $b = (b_{i,j})_{1 \leq i \leq r, 1 \leq j \leq r}$ with coordinates in A such that $ab = I$. Hence $\det a \cdot \det b = 1$ whence a is invertible. Since a is invertible it is immediate that $\{a_1, \ldots, a_r\}$ is a basis for M. This establishes the lemma.

6.5. Proposition. Let A be a ring, M be an A-module locally of finite presentation and r be an integer ≥ 0. The following conditions are equivalent:

(1) There exists $N \geq 1$ and a subset $\{f_1, \ldots, f_N\}$ of A which generates the unit ideal in A such that for each

$1 \leq i \leq N$ M_{f_i} can be generated as an A_{f_i}-module by a subset with $\leq r$ elements.

(2) There exists $N \geq 1$ and a subset $\{f_1, \ldots, f_N\}$ of A which generates the unit ideal in A such that for each $1 \leq i \leq N$ there exist $m, n \geq 0$ and a presentation

$$(*) : A_{f_i}^m \longrightarrow A_{f_i}^n \longrightarrow M_{f_i} \longrightarrow 0 \text{ for } M_{f_i} \text{ as an } A_{f_i}\text{-module}$$

such that $I_r^{A_{f_i},(*)}(M_{f_i}) = A_{f_i}$. (In fact we can insist that $n = r$.)

(2') There exists $N \geq 1$ and a subset $\{f_1, \ldots, f_N\}$ of A which generates the unit ideal in A such that for each $1 \leq i \leq N$ and each presentation

$$(*) : A_{f_i}^m \longrightarrow A_{f_i}^n \longrightarrow M_{f_i} \longrightarrow 0, \; m, n \geq 0 \text{ for } M_{f_i}$$

as an A_{f_i}-module we have $I_r^{A_{f_i},(*)}(M_{f_i}) = A_{f_i}$.

(3) For each $P \in \text{Spec}(A)$ there exist $m, n \geq 0$ and a presentation $(*) : A_P^m \longrightarrow A_P^n \longrightarrow M_P \longrightarrow 0$ for M_P as an A_P-module with $n \leq r$ (in fact we can insist that $n = r$) and for each such presentation $(*)$ we have $I_r^{A_P,(*)}(M_P) = A_P$.

(3') For each $P \in \text{Spec}(A)$ and each presentation

$$(*) : A_P^m \longrightarrow A_P^n \longrightarrow M_P \longrightarrow 0, \; m, n \geq 0 \text{ for } M_P$$

as an A_P-module we have $I_r^{A_P,(*)}(M_P) = A_P$.

Proof.

(1) $\Longrightarrow$ (3). Fix $P \in \text{Spec}(A)$. In view of (1) we conclude that M_P

can be generated as an A_P-module by a subset with $\leq r$ elements. Hence by Lemma 6.3 we conclude that there exists a presentation (*) : $A_P^m \longrightarrow A_P^n \longrightarrow M_P \longrightarrow 0$ for M_P as an A_P-module, $m \geq 0$ with $0 \leq n \leq r$, and more strongly we can choose $n = r$. Hence by construction it is immediate that $I_r^{A_P, (*)}(M_P) = A_P$. This establishes (3).

(3) $\Longrightarrow$ (1). Immediate by Lemma A.14 (in the appendix) and Remark 1.1.1, (1).

(1) & (3) $\Longrightarrow$ (2). Let N and $\{f_1, \ldots, f_N\}$ be as in (1). Fix $1 \leq i_0 \leq N$. Hence to establish (2) at f_{i_0} we may replace $M_{f_{i_0}}$ by M and $A_{f_{i_0}}$ by A. Hence there exists an epimorphism $\psi : A^r \longrightarrow M$ of A-modules. For each $P \in \text{Spec}(A)$ (resp., for each $f \in A$) let $\psi_P : A_P^r \longrightarrow M_P$ (resp., let $\psi_f : A_f^r \longrightarrow M_f$) be the epimorphism of A_P-modules (resp., of A_f-modules) obtained from ψ by passing to quotients. Fix $P \in \text{Spec}(A)$.

Since $\text{Ker}(\psi_P)$ is a finitely generated A_P-module, by Lemma 6.3 there exist $m_P \geq 0$ and a subset $\{a_1', \ldots, a_{m_P}'\}$ of A_P^r which generates $\text{Ker}(\psi_P)$ as an A_P-module. Clear all denominators in all coordinates of each member of the set $\{a_1', \ldots, a_{m_P}'\}$ yielding a subset $\{a_1^{(P)}, \ldots, a_{m_P}^{(P)}\}$ of A^r which generates $\text{Ker}(\psi_P)$ as an A_P-module. By Lemma A.14 (in the appendix) there exists $f_P \in A$ such that $f_P \notin P$ and such that $\{a_1^{(P)}, \ldots, a_{m_P}^{(P)}\}$ generates $\text{Ker}(\psi_{f_P})$ as an A_{f_P}-module. Since $\text{Spec}(A) = \bigcup_{P \in \text{Spec}(A)} D(f_P)$ and since by (1) in Remark 1.1.1 $\text{Spec}(A)$ is quasicompact, there exist $t \geq 1$

and a subset $\{P_1, \ldots, P_t\}$ of Spec(A) such that

(4) $$\text{Spec}(A) = \bigcup_{i=1}^{t} D(f_{P_i}) .$$

For each $1 \le i \le t$ and each $1 \le j \le m_{P_i}$ put $a_{i,j} = a_j^{(P_i)}$, $m_i = m_{P_i}$, $f_i = f_{P_i}$ and $\psi_i = \psi_{P_i}$.

Fix $b \in \text{Ker}(\psi)$. Therefore $b \in \bigcap_{i=1}^{t} \text{Ker}(\psi_i)$. Hence for each $1 \le i \le t$ and $1 \le j \le m_i$ there exist $c'_{i,j} \in A$ and $s_i \ge 1$ such that

(5) $$f_i^{s_i} b = \sum_{j=1}^{m_i} c'_{i,j} a_{i,j} .$$

Put $s = s_1 \ldots s_t$. In view of (4) we obtain $\text{Spec}(A) = \bigcup_{i=1}^{t} D(f_i^{s_i})$. Hence for each $1 \le i \le t$ there exists $g_i \in A$ such that $1 = \sum_{i=1}^{t} f_i^s g_i^s$. For each $1 \le i \le t$ and $1 \le j \le m_i$ put $c_{i,j} = f_i^{s-s_i} g_i^s c'_{i,j}$. In view of (5) we obtain for each $1 \le i \le t$ that $f_i^s g_i^s b = \sum_{j=1}^{m_i} c_{i,j} a_{i,j}$. Hence $b = (\sum_{i=1}^{t} f_i^s g_i^s) b = \sum_{i=1}^{t} f_i^s g_i^s b = \sum_{i=1}^{t} (\sum_{j=1}^{m_i} c_{i,j} a_{i,j})$. This shows that $\{a_{i,j} | 1 \le i \le t, 1 \le j \le m_i\}$ generates $\text{Ker}(\psi)$ as an A-module. Since we can choose $m = t(m_1 + \ldots + m_t)$ this proves the existence of a presentation $(*): A^m \longrightarrow A^r \longrightarrow M \longrightarrow 0$ for M as an A-module. It is now immediate by construction that $I_r^{A,(*)}(M) = A$.

(2) for $n = r \Longrightarrow$ (1). Obvious.

(3') $\Longrightarrow$ (2'). Since M is an A-module locally of finite presentation, in view of (3') and Cor. 6.2.1 we obtain (2').

(2') $\Longrightarrow$ (2). Immediate since M is an A-module locally of finite presentation.

(3) $\Longrightarrow$ (3'). Fix $P \in \text{Spec}(A)$ and let (*) be as in (3'). If $n \leq r$, by (3) we obtain the conclusion of (3'). Hence we may suppose $n > r$. Hence by Cor. 6.3.1 there exists a presentation

(**) : $A_P^m \longrightarrow A_P^r \longrightarrow M \longrightarrow 0$, $m \geq 0$ for M_P as an A_P-module such that

(6) $\qquad I_i^{A_P,(*)}(M_P) = I_i^{A_P,(**)}(M_P)$ for each $i \geq 0$.

But by (3) we conclude

(7) $\qquad I_r^{A_P,(*)}(M_P) = A_P$.

In view of (6) and (7) we obtain the conclusion of (3').

This establishes the proposition.

6.5.1. Corollary. Let A be a ring, M be an A-module locally of finite presentation and r be an integer ≥ 0. The following conditions are equivalent:

(1) M is a locally free A-module of rank r.

(2) There exists $N \geq 1$ and a subset $\{f_1, \ldots, f_N\}$ of A which generates the unit ideal in A such that for each $1 \leq i \leq N$ there exist $m, n \geq 0$ and a presentation (*) : $A_{f_i}^m \longrightarrow A_{f_i}^n \longrightarrow M_{f_i} \longrightarrow 0$ for M_{f_i} as an A_{f_i}-module such that

$I_r^{A_{f_i},(*)}(M_{f_i}) = A_{f_i}$ and $I_{r-1}^{A_{f_i},(*)}(M_{f_i}) = 0$. (In fact we can insist that $n = r$ and $m = 0$.)

(2') There exists $N \geq 1$ and a subset $\{f_1, \ldots, f_N\}$ of A which generates the unit ideal in A such that for each $1 \leq i \leq N$ and each presentation

$(*): A_{f_i}^m \longrightarrow A_{f_i}^n \longrightarrow M_{f_i} \longrightarrow 0, \; m, n \geq 0$ for M_{f_i}

as an A_{f_i}-module we have $I_r^{A_{f_i},(*)}(M_{f_i}) = A_{f_i}$ and $I_{r-1}^{A_{f_i},(*)}(M_{f_i}) = 0$.

(3) For each $P \in \text{Spec}(A)$ we have an isomorphism $A_P^r \longrightarrow M_P$ of A_P-modules yielding a presentation

$(*): 0 \longrightarrow A_P^r \longrightarrow M_P \longrightarrow 0$ for M_P as an A_P-module. Moreover, $I_r^{A_P,(*)}(M_P) = A_P$ and $I_{r-1}^{A_P,(*)}(M_P) = 0$.

(3') For each $P \in \text{Spec}(A)$ and each presentation

$(*): A_P^m \longrightarrow A_P^n \longrightarrow M_P \longrightarrow 0, \; m, n \geq 0$ for M_P

as an A_P-module we have $I_r^{A_P,(*)}(M_P) = A_P$ and $I_{r-1}^{A_P,(*)}(M_P) = 0$.

Proof.

(1) $\Longrightarrow$ (3). Fix $P \in \text{Spec}(A)$. In view of (1) we obtain an isomorphism $A_P^r \longrightarrow M_P$ of A_P-modules. Hence by Prop. 6.5 we conclude $I_r^{A_P,(*)}(M_P) = 0$ where $(*)$ is the presentation $0 \longrightarrow A_P^r \xrightarrow{\sim} M_P \longrightarrow 0$. Also it is immediate by construction that

$I_{r-1}^{A_P,(*)}(M_P) = 0$. This establishes (3).

(3) $\Longrightarrow$ (1). Immediate by Cor. A.11.2 (in the appendix).

(1) & (3) $\Longrightarrow$ (2). In view of (1) and Remark 1.1.1, (1) there exists $N \geq 1$ and a subset $\{f_1, \ldots, f_N\}$ of A which generates the unit ideal in A and an isomorphism $A_f^r \longrightarrow M_{f_i}$ of A_{f_i}-modules for each $1 \leq i \leq N$. Fix $1 \leq i \leq N$ and let (*) be the presentation $0 \longrightarrow A_{f_i}^r \xrightarrow{\approx} M_{f_i} \longrightarrow 0$ for M_{f_i} as an A_{f_i}-module. Hence by construction we conclude

$I_r^{A_{f_i},(*)}(M_{f_i}) = A_{f_i}$ and $I_{r-1}^{A_{f_i},(*)}(M_{f_i}) = 0$. This establishes (2).

(2) for $n = r$ & $m = 0$ $\Longrightarrow$ (1). Obvious.

(3') $\Longrightarrow$ (2'). Immediate by Cor. 6.2.1 since M is an A-module locally of finite presentation.

(2') $\Longrightarrow$ (2). Immediate since M is an A-module locally of finite presentation.

(1) & (3) $\Longrightarrow$ (3'). Fix $P \in \text{Spec}(A)$ and let (*) be as in (3'). Note $n \geq r$. By Cor. 6.3.1 there exists a presentation

(**) : $A_P^m \longrightarrow A_P^r \longrightarrow M_P \longrightarrow 0$, $m \geq 0$ for M_P as an A_P-module such that

(4) $\qquad I_i^{A_P,(*)}(M_P) = I_i^{A_P,(**)}(M_P)$ for all $i \geq 0$.

But by (1) and Lemma 6.4 we conclude that the homomorphism $A_P^m \longrightarrow A_P^r$ in the presentation (**) is the zero homomorphism. Hence by construction we obtain

(5) $$I_{r-1}^{A_P,(**)}(M_P) = 0 .$$

In view of (4) and (5) we obtain $I_{r-1}^{A_P,(*)}(M_P) = 0$. By (3) and Prop. 6.5 we obtain $I_r^{A_P,(*)}(M_P) = A_P$. This establishes (3').

This establishes the corollary.

6.6. Theorem. Let A be a ring and B be a finitely presented A-algebra. Hence there exists $n \geq 1$ and a finitely generated ideal I in $A[T_1, \ldots, T_n]$ such that B canonically identifies to $A[T_1, \ldots, T_n]/I$ as an A-algebra. Let $\{f_1, \ldots, f_m\}$ be a set of generators for the ideal I in $A[T_1, \ldots, T_n]$.

In view of Cor. 2.10.2 we obtain a presentation

(*) : $B^m \xrightarrow{g} B^n \xrightarrow{d_{B/A}} \Gamma_A^1(B)$ for $\Gamma_A^1(B)$ as a B-module, where the matrix a corresponding to the homomorphism g with respect to the standard bases is $(\partial f_i / \partial T_j)_{1 \leq i \leq m, 1 \leq j \leq n}$.

Let r be an integer ≥ 0. The following are true:

(1) The following three conditions are equivalent:

(1A) There exists $N \geq 1$ and a subset $\{b_1, \ldots, b_N\}$ of B which generates the unit ideal in B such that for each $1 \leq i \leq N$ $\Gamma_A^1(B)_{b_i} = \Gamma_A^1(B_{b_i})$ can be generated as a B_{b_i}-module by a subset with $\leq r$ elements.

(1B) For each $P \in \mathrm{Spec}(B)$, $\Gamma_A^1(B)_P = \Gamma_A^1(B_P)$ can be generated as a B_P-module by a subset with $\leq r$ elements.

(1C) $I_r^{B,(*)}(\Gamma_A^1(B)) = B$.

(2) The following two conditions are equivalent:

(2A) $\Gamma_A^1(B)$ is a locally free B-module of rank r.

(2B) $I_r^{B,(*)}(\Gamma_A^1(B)) = B$ and $I_{r-1}^{B,(*)}(\Gamma_A^1(B)) = 0$.

Proof.

(1) is immediate by Prop. 6.5 and (2) is immediate by Cor. 6.5.1.

6.6.1. Corollary. With notation and hypotheses as in Thm. 6.6 for each $P \in \text{Spec}(B)$ denote by $a(P)$ the matrix defined in Def. 1.2. Let r be an integer ≥ 0. Then the following condition is equivalent to the three equivalent conditions in (1) in Th. 6.6:

(1D) $\text{rank}_{\kappa(P)} a(P) \geq n-r$ for each $P \in \text{Spec}(B)$.

Proof.

For each $P \in \text{Spec}(B)$ let $(*_P) : B_P^m \longrightarrow B_P^n \longrightarrow \Gamma_A^1(B)_P$ be the presentation for $\Gamma_A^1(B)_P$ as a B_P-module obtained from (*) by passing to quotients. By Prop. 6.5 condition (1C) holds if and only if for each $P \in \text{Spec}(B)$

(3) $\qquad I_r^{B_P,(*_P)}(\Gamma_A^1(B)_P) = B_P$.

By construction it is immediate that (3) holds at P if and only if at least one n-r minor of $a(P)$ is nonzero which is true if and only if $\text{rank}_{\kappa(P)} a(P) \geq n-r$. This establishes the corollary.

CHAPTER 7

Proof of the Jacobian criterion and some characterizations of simple k-algebras and A-algebras

In this chapter many important properties and characterizations of simple k-algebras and A-algebras are presented. In particular, are Thm. 7.1 which establishes the Jacobian criterion and Thm. 7.8 which generalizes Thm. 3.5. Other results in this chapter of major interest and importance are Cor. 7.1.1.N (The noetherian hypotheses in Cor. 7.1.1.N are removed in Chapter 9.), Thm. 7.5 and Cor. 7.5.1.

7.1. **Theorem.** Let k be a field, B be a k-algebra, $Q \in \operatorname{Spec}(B)$, put $A = B_Q$ and $n = \sup_P \operatorname{tr. deg.} \kappa(P)/k$, where P runs through the set of minimal prime ideals in A. (See Remarks 1.3.1, (2) & (3) for alternate characterizations of n.) Suppose there exists $f \in B$ such that $Q \in D(f)$ and such that B_f is a finitely generated k-algebra. Then the following conditions are equivalent:

(1). A is simple over k.

(2). The minimum cardinality of a set of generators for $\Gamma_k^1(A)$ as an A-module $= n$.

(3). $\Gamma_k^1(A)$ is a free A-module of rank n.

(4). There exists a subset $\{x_1, \ldots, x_n\}$ of A which is algebraically independent over k such that $\Gamma_k^1(A)$ is a free A-module of rank n with basis $\{d(x_1), \ldots, d(x_n)\}$.

Since B_f is a finitely generated k-algebra there exists an integer $N \geq 1$ and a finitely generated ideal I of $k[T_1, \ldots, T_N]$ such that B_f identifies as k-algebra to $k[T_1, \ldots, T_N]/I$. Let $M \geq 1$ and $\{f_1, \ldots, f_M\}$ be a set of generators for the ideal I in $k[T_1, \ldots, T_N]$. Let a denote the Jacobian matrix $(\partial f_i / \partial T_j)_{1 \leq i \leq M, 1 \leq j \leq N}$ and for each $P \in D(f)$ let $a(P)$ be the matrix defined in Def. 1.2. Let K denote the residue class field of A. Then the following two conditions are equivalent to the four conditions above:

(5). $\text{rank}_K a(Q) \geq N - n$.

(6). $\text{rank}_K a(Q) = N - n$.

Proof.

(1) $\Rightarrow$ (2). Immediate by Thm. 3.5.

(2) $\Rightarrow$ (3). Immediate by Cor. 5.2.

(3) $\Rightarrow$ (4). Immediate by Thm. 3.5 & Cor. 5.4.

(4) $\Rightarrow$ (1). Obvious.

(1) $\Leftrightarrow$ (5). Immediate by Cor. 6.6.1 and Prop. A.16 (in the appendix).

(2) $\Leftrightarrow$ (6). Immediate by Cor. 6.6.1 and Prop. A.16 (in the appendix).

This establishes the theorem.

Conditions (5) and (6) of Thm. 7.1 are the <u>Jacobian criterion</u> which in Chapter 1 was used as the definition of a simple k-algebra A.

7.1.1. Corollary. Let A be a ring, B be an A-algebra and let $Q \in \text{Spec}(B)$. Suppose B is simple over A at Q. (See Def. 7.2 below.) Then there exists an open neighborhood U of Q in Spec (B) such that B is simple over A at P for all P ∈ U.

Proof.

We postpone the proof until 9.3 in Chapter 9.

7.1.1.N. Corollary. With hypotheses and notation as in Cor. 7.1.1 we assume, in addition, that there exists $f \in B$ such that $Q \in D(f)$ and such that B_f has only finitely many minimal prime ideals.

(Note that this added hypothesis is automatically satisfied whenever the ring B is noetherian and thus also automatically satisfied whenever the ring A is noetherian since B being simple over A at Q implies we can also insist that f be so chosen that B_f is a finitely presented A-algebra whence the ring B_f would be noetherian.)

Proof.

Let $\lambda : A \to B$ be the structure homomorphism, put $P = \lambda^{-1}(Q)$, let Q' be the image of Q in $B \otimes_A \kappa(P)$, and put n = sup tr. deg. $\kappa(R)/\kappa(P)$, where R runs through the set of minimal prime ideals in $(B \otimes_A \kappa(P))_{Q'}$. By choosing f as in the note in the statement of the corollary and by replacing B by B_f, we can assume without loss of generality that B is a finitely presented A-algebra and that B has only finitely many minimal prime ideals. Put S = {R|R is a minimal prime ideal in B and tr. deg. $\kappa(R)/\kappa(\lambda^{-1}(R)) \dagger n$}. By hypothesis the set S is finite.

For each $R \in S$ choose $f_R \in R \setminus Q$ and put $f_0 = \prod_{R \in S} f_R$. Note $f_0 \in (\bigcap_{R \in S} R) \setminus Q$. Hence given a minimal prime ideal $R \in D(f_0)$ we have

(1) $\qquad$ tr. deg. $\kappa(R)/\kappa(\lambda^{-1}(R)) = n$.

Note that by construction we have $Q \in D(f_0)$.

Since B is a finitely presented A-algebra there exists an integer $N \geq 1$ and a finitely generated ideal I of $A[T_1, \ldots, T_N]$ such that $B = A[T_1, \ldots, T_N]/I$. Let $M \geq 1$ and $\{f_1, \ldots, f_M\}$ be a set of generators for the ideal I. Let a denote the Jacobian matrix $(\partial f_i/\partial T_j)_{1 \leq i \leq M, 1 \leq j \leq N}$ and for each $R \in \text{Spec}(B)$ denote by $a(R)$ the matrix defined in Def. 1.2.

Since B is simple over A at Q, by the Jacobian criterion (condition (5) in Thm. 7.1) we conclude

(2) $\qquad \text{rank}_{\kappa(Q)} a(Q) \geq N - n$.

In view of (2) we conclude that the matrix $a(Q)$ has a nonzero $(N-n) \times (N-n)$ minor. Hence by Prop. A.16 (in the appendix) we conclude that there exists $g \in B$ such that the matrix $a(R)$ has a nonzero $(N-n) \times (N-n)$ minor for each $R \in D(g)$ whence we conclude

(3) $\qquad \text{rank}_{\kappa(R)} a(R) \geq N - n$ for each $R \in D(g)$.

B_Q is a flat A_P-module since B is simple over A at Q. Hence by Thm. A. (in the appendix) we conclude

(4) $\begin{cases} \text{There exists } h \in B \text{ such that } B_R \text{ is a flat } A_{\lambda^{-1}(R)} \text{-module} \\ \text{for each } R \in D(h). \end{cases}$

Put $f = f_0 gh$. In view of (4), by (1) and (3) and the Jacobian criterion (condition (5) in Thm. 7.1) we conclude that B_f is simple over A. This establishes the corollary.

7.1.1.1. **Corollary.** With notation and hypotheses as in Thm. 7.1 suppose B is simple over k at Q. Then there exists an open neighborhood U of Q in Spec (B) such that B is simple over k at P for all $\dot{P} \in U$.

Proof.

Immediate by Cor. 7.1.1.N since B_f, being a finitely generated k-algebra, is noetherian.

7.2. **Definition.** Let A be a ring, B be an A-algebra, $\lambda : A \to B$ be the structure homomorphism, $Q \in \text{Spec } (B)$, $P = \lambda^{-1}(Q)$ and let Q' denote the image of Q in $B \otimes_A \kappa(P) = B/\mathcal{M}_{A_P} B$. We say that <u>B is simple over A at Q</u> or <u>λ is simple at Q</u> if and only if $B \otimes_A \kappa(P)$ is simple over $\kappa(P)$ at Q', B_Q is a flat A_P-module and there exists $f \in B$ such that $f \notin Q$ and B_f is a finitely presented A-algebra. We say that <u>B is simple over A</u> or <u>B is a simple A-algebra</u> or <u>λ is simple</u> if and only if B is simple over A at Q for all $Q \in \text{Spec } (B)$.

7.2.1. Remark. The condition "there exists $f \in B$ such that $f \notin Q$ and B_f is a finitely presented A-algebra" in Def. 7.2 is not superfluous as the example in Remark 4.2.3 illustrates.

7.2.2. Remark. Let A be a ring and B be an A-algebra. If B is simple over A then B is a finitely presented A-algebra; if B is étale over A then B is simple over A. (Both of these facts are immediate by definition.)

7.3. Lemma. Let k be a field, B be a k-algebra and let $Q \in \text{Spec}(B)$. Suppose B is simple over k at Q. Then by Cor. 7.1.1.1 there exists $f \in B$ such that $Q \in D(f)$ and such that B is simple over k at P for all $P \in D(f)$. Then the following are true:

(1). $\Gamma_k^1(B)_f$ is a locally free B_f-module of finite rank. In particular, for each $P \in D(f)$ we have $\Gamma_k^1(B)_P$ is a free B_P-module of rank $\sup_R \text{tr. deg. } \kappa(R)/k$, where R runs through the set of minimal elements in the set of prime ideals of B contained in P.

(2). For each minimal prime ideal P in B_f (that is for each prime ideal P of B not containing f minimal in the set of prime ideals of B not containing f) we have $B_P = \kappa(P)$ and $\kappa(P)$ is a separable field extension of k.

Proof.

Conclusion (1) of the lemma is immediate by condition (3) in Thm. 7.1 and Cor. 7.1.1.1. Let P be as in conclusion (2) of the lemma. Hence B_P, being a noetherian local ring whose only prime ideal is the maximal ideal, is artinian.

But B_P, being simple over k by hypothesis, is an integral domain by Thm. 3.5. Hence B_P, being an artinian local domain, is a field, namely the field $\kappa(P)$. Hence by Cor. 5.3 we conclude that $\kappa(P)$ is a separable field extension of k, which establishes (2).

7.4. Lemma. Let k be a field, B be a k-algebra and let $f \in B$ such that B_f is a finitely generated k-algebra, such that the first statement in (1) of Lemma 7.3 holds and such that (2) of Lemma 7.3 holds. Then B is simple over k at P for all $P \in D(f)$.

Proof.

By replacing B by B_f we may assume without loss of generality that B is a finitely generated k-algebra. Let $f \in B$ be as in (1) & (2) of Lemma 7.3 and let $Q \in D(f)$. Since $\Gamma_k^1(B)$ is a finitely presented B-module (by Cor. 2.10.2), by the first statement in (1) of Lemma 7.3 and Cor. A.11.2 (the latter in the appendix) we conclude that there exists $g \in B$ such that $Q \in D(g)$ and such that $\Gamma_k^1(B_Q)$ is a free B_Q-module of finite rank n equal to the rank of the free B_{fg}-module $\Gamma_k^1(B_{fg})$.

Since it suffices to show that B is simple over k at P for all $P \in D(fg)$, without loss of generality we may replace f by fg. Now suppose Q is a minimal prime ideal in B_f. Since $\kappa(Q)$ is a separable field extension of k by (2) in Lemma 7.3, by Prop. 3.2 we conclude that $\Gamma_k^1(\kappa(Q))$ is a $\kappa(Q)$-vector space of dimension tr. deg. $\kappa(Q)/k$. Hence by Nakayama's lemma we conclude n = tr. deg. $\kappa(Q)/k$.

Hence we have shown for each $Q \in D(f)$ that $\Gamma^1_k(B_Q)$ is a free B_Q-module of rank $n = \text{tr. deg. } \kappa(P)/k$, where P is any minimal prime ideal in B_f. Hence by condition (3) in Thm. 7.1 we conclude that B is simple over k at P for all $P \in D(f)$. This establishes the lemma.

7.5. Theorem. Let k be a field and B be a k-algebra. Then the following two conditions are equivalent:

(1). B is simple over k.

(2). The following three conditions hold:

(2A). B is a finitely generated k-algebra.

(2B). $\Gamma^1_k(B)$ is a locally free B-module.

(2C). For each minimal prime ideal P in B we have that $\kappa(P)$ is a separable field extension of k.

Proof.

Immediate by Lemma 7.3 and Lemma 7.4.

7.5.1. Corollary. Let A be a ring and B be an A-algebra with structure homomorphism $\lambda : A \to B$. Then the following two conditions are equivalent:

(1). B is simple over A.

(2). The following four conditions hold:

(2A). B is a finitely presented A-algebra.

(2B). $\Gamma^1_A(B)$ is a locally free B-module.

(2C). For each $Q \in \text{Spec}(B)$, putting $P = \lambda^{-1}(Q)$, and each minimal element R in the set of prime ideals in B

lying over P, we have $\kappa(R)$ is a separable field extension of $\kappa(P)$.

(2D). For each $Q \in \text{Spec}(B)$, putting $P = \lambda^{-1}(Q)$, B_Q is a flat A_P-module (or equivalently, B is a flat A-module).

Proof.

(1) $\Rightarrow$ (2). Note that (2A) is immediate by (1) by definition. By (1) and Thm. 7.5 we obtain (2C) and (2D). We proceed to establish (2B) under the additional hypothesis that the ring A is reduced. We postpone the proof of (1) $\Rightarrow$ (2B) for general A until just after the proof of Thm. 8.2.N in the next chapter. There will be no vicious circle of proofs resulting from delaying the proof of (1) $\Rightarrow$ (2B) for general A since we shall only use this corollary to establish Thm. 8.2.N and then only under the added hypothesis that the ring A is reduced.

Note that $\Gamma_A^1(B)$ is a finitely presented B-module by (2A) and Cor. 2.10.2. By definition and Thm. 7.5 we conclude that for each $P \in \text{Spec}(A)$, $\Gamma_{\kappa(P)}^1(B \otimes_A \kappa(P)) = \Gamma_A^1(B) \otimes_A \kappa(P)$ is a locally free $B \otimes_A \kappa(P)$-module of say rank n_P, and since the rank is locally constant by Cor. A.11.2 in the appendix (that is for each $P \in \text{Spec}(A)$ there exists $f \in A$ such that $P \in D(f)$ and such that $n_{P_1} = n_{P_2}$ for all $P_1, P_2 \in D(f)$), by Cor. A.12.1 in the appendix we conclude that $\Gamma_A^1(B)$ is a locally free A-module. Since B is a flat finitely presented A-algebra by (1) and (2A), since $\Gamma_A^1(B)$ is a finitely presented B-module and since $\Gamma_A^1(B)$ is a flat A-module, by Thm. A.10 in the appendix we conclude that $\Gamma_A^1(B)$ is a flat B-module. $\Gamma_A^1(B)$, being a finitely presented flat B-module, is a locally free B-module. This establishes (2B) for the case where the ring A is reduced.

(2) $\Rightarrow$ (1). Immediate by definition and Thm. 7.5.

7.5.2. Corollary. Let k be a field and n be an integer ≥ 1. Then $k[T_1, \ldots, T_n]$ is simple over k.

Proof.

Immediate by Example 1.1 and Thm. 7.5.

7.6. Corollary. Let k be a field, B be a k-algebra, $Q \in \text{Spec}(B)$, put $A = B_Q$ and let K denote the residue class field of A. Suppose there exists $f \in B$ such that $Q \in D(f)$ and such that B_f is a finitely generated k-algebra and suppose that K is a separable field extension of k. Then the following two conditions are equivalent:

    (1). A is simple over k.

    (2). A is a regular local ring.

Proof.

(1) $\Rightarrow$ (2). Immediate by Lemma 3.4.

(2) => (1). Let $\mathfrak{m}$ denote the maximal ideal of A and L denote the quotient field of A, put m = tr. deg. K/k, n = tr. deg. L/k, put r = dim A and let n_0 denote the minimum cardinality of a set of generators for $\Gamma^1_k(A)$ as an A-module.

By (3) in Thm. A.6 (in the appendix) we have

(3) $r + m = n.$

Since A contains the field k and since K is a separable field extension of k by hypothesis, by Prop. 3.1 we have an exact sequence of K-vector spaces

(4) $$0 \to m/m^2 \to \Gamma_k^1(A) \otimes_A K \to \Gamma_k^1(K) \to 0.$$

Since A is a regular local ring we have

(5) $$\dim_K(m/m^2) = r.$$

Since K is a separable field extension of k, by Prop. 3.2 we conclude

(6) $$\dim_K(\Gamma_k^1(K)) = m.$$

In view of (3), (4), (5) and (6) we obtain $n_0 = n$. Hence by definition we obtain (1).

7.6.1. Corollary. Let k be a perfect field, B be a finitely generated k-algebra, $Q \in \text{Spec}(B)$ and put $A = B_Q$. Then A is simple over k if and only if A is a regular local ring.

Proof.

Immediate by Cor. 7.6.

7.7. Lemma. Let k be a field, B be a finitely generated k-algebra, $Q \in \text{Spec}(B)$ and put $A = B_Q$. The following conditions are equivalent:

(1). A is simple over k.

(2). For each purely inseparable algebraic field extension k', of k, the local ring $A \otimes_k k'$ is regular.

(3). The local ring $A \otimes_k k^{p^{-\infty}}$ is regular, where p is the characteristic exponent of k.

Proof.

(1) $\Rightarrow$ (2). Immediate by the proof of Thm. 3.5.

(2) $\Rightarrow$ (3). Obvious.

(3) $\Rightarrow$ (1). If k is of characteristic 0, by Cor. 7.6 we obtain (1). Hence we may assume k is of characteristic $p > 0$. Put $k' = k^{p^{-\infty}}$. Let B', Q' and A' be as in the proof of Thm. 3.5. Note by the proof of Thm. 3.5 that B' is integral over B and the natural homomorphism $B \to B'$ is injective. Put $A_0 = A \otimes_k k' = B_Q \otimes_k k'$. Let Q_0 be the maximal ideal of A_0. Then $A' = A_{0Q'}$. Hence by (3) we conclude that A' is a regular local ring.

Since the field k' is perfect, by Cor. 7.6.1 we conclude that A' is simple over k'. Hence by Cor. 7.1.1.N we conclude that there exists $f \in B'$ such that $Q' \in D(f)$ and such that B'_f is simple over k'. Note that $f^{p^r} \in B$ for some integer $r \geq 0$. Since $D(f) = D(f^{p^r})$ (as subsets of Spec (B')), by replacing f by f^{p^r} we may assume without loss of generality that $f \in B$. Hence by Thm. 7.5 we conclude

(3) $\qquad \Gamma^1_{k'}(B'_f)$ is a locally free B'_f-module.

But

(4) $$\begin{cases} \Gamma^1_{k'}(B'_f) = \Gamma^1_{k'}(B')_f = \Gamma^1_{k'}(B \underset{k}{\otimes} k')_f = \\ = (\Gamma^1_k(B) \underset{k}{\otimes} k')_f = \Gamma^1_k(B_f) \underset{k}{\otimes} k' . \end{cases}$$

Since $B'_f = B_f \underset{k}{\otimes} k'$ is a faithfully flat B_f-module, in view of (3) & (4), by Prop. A.13 (in the appendix) we conclude

(5) $\quad\quad\quad \Gamma^1_k(B_f)$ is a locally free B_f-module.

Let P be a minimal prime ideal in B_f. B'_f is integral over B_f and the homomorphism $B_f \to B'_f$ is injective since B' is integral over B and the homomorphism $B \to B'$ is injective. Hence there exists a prime ideal P' in B'_f lying over P. Since P is a minimal prime ideal in B_f and B'_f is integral over B_f we conclude that P' is the only prime ideal in B'_f lying over P and P' is a minimal prime ideal in B'_f.

Let n be as in the proof of Lemma 3.4 and n' be as in the proof of Thm. 3.5. By the proof of Lemma 7.4 we have

(6) $$\begin{cases} \Gamma^1_{k'}(B'_{P'}) \text{ is a free } B'_{P'}\text{-module} \\ \text{of rank } n' \text{ and } n' = \text{tr. deg. } \kappa(P')/k' . \end{cases}$$

In view of (5) we have $\Gamma^1_k(B_P)$ is a free B_P-module. Since $B'_{P'} = B'_P = (B \otimes_k k')_P = B_P \underset{k}{\otimes} k'$ we have $\Gamma^1_{k'}(B'_{P'}) = \Gamma^1_{k'}(B_P \underset{k}{\otimes} k') = \Gamma^1_k(B_P) \underset{k}{\otimes} k'$ whence in view of (6) we conclude

(7) $\quad\quad\quad \Gamma^1_k(B_P)$ is a free B_P-module of rank n'.

Since the prime ideals in B'_f lying over P are in one-to-one correspondence with the prime ideals in $B'_f \otimes_k \kappa(P)$ and since P' is the only prime ideal in B'_f lying over P we conclude

(8) $\begin{cases} B' \otimes_k \kappa(P) \text{ is a local ring with residue class field} \\ \kappa(P'). \end{cases}$

$B' \otimes_k \kappa(P)$ is a finitely generated $\kappa(P) \otimes_k k'$-algebra since B' is a finitely generated k'-algebra. Hence

(9) $\quad B' \otimes_k \kappa(P)$ is a finitely generated $\kappa(P)(k')$-algebra.

In view of (8), (9) and Prop. A.15 (the latter in the appendix) we conclude that $\kappa(P')$ is a finite field extension of $\kappa(P)(k')$. But tr. deg. $\kappa(P)(k')/k'$ = tr. deg. $\kappa(P)/k$ since k' is algebraic over k. Hence tr. deg. $\kappa(P')/k'$ = tr. deg. $\kappa(P)/k$ whence by definition of n and n' we have

(10) $\quad\quad\quad\quad\quad\quad\quad\quad n = n'$.

In view of (7) and (10) we obtain

(11) $\quad\quad\quad\quad \Gamma^1_k(B_P)$ is a free B_P-module of rank n.

By Prop. 2.9 we have an epimorphism $\Gamma^1_k(B_P) \otimes_{B_P} \kappa(P) \to \Gamma^1_k(\kappa(P))$ of $\kappa(P)$-vec spaces whence in view of (11) we conclude

(12) $$\begin{cases} \Gamma^1_k(\kappa(P)) \text{ is a } \kappa(P)\text{-vector space} \\ \text{of dimension } \leq n. \end{cases}$$

Hence in view of (12), by Prop. 3.2 we conclude

(13) $\kappa(P)$ is a separable field extension of k.

In view of (5) and (13), by Thm. 7.5 we conclude that B_f is simple over k whence by definition we obtain (1).

7.8. **Theorem.** Let k be a field, B be a k-algebra, $Q \in \text{Spec}(B)$ and put $A = B_Q$. Suppose there exists $f \in B$ such that $Q \in D(f)$ and such that B_f is a finitely generated k-algebra. Then the following conditions are equivalent:

(1). A is simple over k.

(2). For each purely inseparable algebraic field extension k' of k, the local ring $A \otimes_k k'$ is regular.

(3). For each finite purely inseparable algebraic field extension k' of k, the local ring $A \otimes_k k'$ is regular.

(4). For each extension $k' = k^{p^{-s}}$ of k where p is the characteristic exponent of k and s is an integer > 0, the local ring $A \otimes_k k'$ is regular.

(5). The local ring $A \otimes_k k^{p^{-\infty}}$ is regular, where p is the characteristic exponent of k.

(6). For each perfect field extension k' of k, all the local rings of $A \otimes_k k'$ are regular.

(7). For each finite field extension k' of k, all the local rings of $A \otimes_k k'$ are regular.

(8). For each field extension k' of k such that k' is a finitely generated k-algebra, all the local rings of $A \otimes_k k'$ are regular.

(9). For each field extension k' of k, all the local rings of $A \otimes_k k'$ are regular.

Proof.

Immediate by Lemma 7.7 and Prop. A.17 in the appendix.

7.8.1. Remark. We summarize the behavior of the local ring A in conditions (2)-(9) of Thm. 7.8 by saying that the local ring A is geometrically regular.

CHAPTER 8

Characterizations of simple A-algebras in terms of étale homomorphisms; invariance of the property of being a simple algebra under composition and change of base

In this chapter we prove Thm. 8.2.N which gives a characterization of simple A-algebras in terms of étale homomorphisms. With this characterization of simple homomorphisms we are able to prove Cor. 8.2.1.N and Cor. 8.2.3.N which show that simple homomorphisms are invariant under composition and change of base. These two properties were not as evident with the characterizations of simple homomorphisms given through Chapter 7. Also of importance is Cor. 8.2.5. The noetherian assumptions in Thm. 8.2.N, Cor. 8.2.1.N, Cor. 8.2.3.N and various other corollaries in this chapter were needed only because they were needed to prove Cor. 7.1.1.N. In Chapter 9 we shall prove Cor. 7.1.1 which will enable us to remove all noetherian assumptions in this chapter.

The characterization of simple A-algebras given in Thm. 8.2.N (or Thm. 8.2) is taken as a definition of simple A-algebras by some authors, for example, Grothendieck; its notable advantages include a ready proof of Cor. 8.2.1 and Cor. 8.2.3 and simplified proofs of many theorems on simple homomorphisms not treated in this text; some of its disadvantages include complicating the concept of simple homomorphism with the related but different concept of étale homomorphism, its being so far from the classical Jacobian criterion that a proof of the latter is not readily obtained, and its not being a feasible criterion for determining whether a particular example is simple. (Witness the power of the Jacobian criterion applied to the examples in Chapter 1.)

8.1. Lemma. With notation and hypotheses as in Thm. 7.1 suppose A is simple over k. Then by Thm. 7.1 we know $\Gamma^1_k(A) = \Gamma^1_k(B)_Q$ is a free $A = B_Q$-module of rank n and that there exists a subset $\{x_1, \ldots, x_n\}$ of A (and which is algebraically independent over k) such that $\{d(x_1), \ldots, d(x_n)\}$ generates (and is in fact a basis for) the free A-module $\Gamma^1_k(A)$.

For each $1 \leq i \leq n$ choose $b_i \in B$ and $s_i \in B \setminus Q$ such that $x_i = b_i/s_i$ and put $g = s_1 \ldots s_n f$. Hence $\{x_1, \ldots, x_n\} \subset B_g$. Define a homomorphism of k-algebras $\mu : k[T_1, \ldots, T_n] \to B_g$ by $\mu(T_i) = x_i$ for $1 \leq i \leq n$. Put $Q' = QB_g$. Then B_g is étale over $k[T_1, \ldots, T_n]$ at Q'.

Proof.

Put $C = k[T_1, \ldots, T_n]$. Note that B_g is a finitely presented C-algebra since B_f is a finitely generated k-algebra. By Prop. 2.9 we have an epimorphism $\Gamma^1_k(A) \to \Gamma^1_C(A)$ of A-modules. Hence since $\{d(x_1), \ldots, d(x_n)\}$ is a basis for the free A-module $\Gamma^1_k(A)$ we conclude that $\{d(x_1), \ldots, d(x_n)\}$ generates the A-module $\Gamma^1_C(A)$. But $d(x_i) = 0$ in $\Gamma^1_C(A)$ for $1 \leq i \leq n$ since $\mu(T_i) = x_i$ for $1 \leq i \leq n$. Hence $\Gamma^1_C(A) = 0$. Hence by Prop. 4.3 we conclude that B_g is unramified over C at Q'.

Put $R = \mu^{-1}(Q')$ and $R' = RC_R$. Since B_g is unramified over C at Q', by Prop. 4.3 we conclude that $R'A = \mathcal{M}_A$ and that K is a finite separable field extension of $\kappa(R)$. Hence $A \otimes_{C_R} \kappa(R) = A/R'A = A/\mathcal{M}_A = K$ whence

(1) $$\dim (A \otimes_{C_R} \kappa(R)) = 0.$$

By Thm. A.6 in the appendix we have

(2) $$\dim A = n - \text{tr. deg. } K/k$$

and

(3) $$\dim C_R = n - \text{tr. deg. } \kappa(R)/k.$$

But since K is algebraic over $\kappa(R)$ we conclude that tr. deg. K/k = tr. deg. $\kappa(R)/k$ which in view of (2) and (3) yields

(4) $$\dim A = \dim C_R.$$

In view of (1) and (4) we obtain

(5) $$\dim A = \dim C_R + \dim(A \otimes_{C_R} \kappa(R)).$$

C is simple over k at R by Cor. 7.5.2 and A is simple over k by hypothesis. Hence by Thm. 3.5 and Thm. 7.1 we conclude that both A and C_R are regular local rings. C_R is Cohen-Macauley, being a regular local ring. Hence in view of (5) we conclude by Prop. A.9 in the appendix that A is a flat C_R-module. Hence by definition B_g is étale over C at Q'. This establishes the lemma.

8.2. Theorem. Let A be a ring, B be an A-algebra with structure homomorphism $\lambda : A \to B$, let $Q \in \mathrm{Spec}(B)$ and put $P = \lambda^{-1}(Q)$. Then the following conditions are equivalent:

(1). B is simple over A at Q.

(2). There exists $f \in B$ such that $Q \in D(f)$, an integer $n \geq 0$ and an étale homomorphism $\mu : A[T_1, \ldots, T_n] \to B_f$ of A-algebras.

(Remark: In condition (2) we necessarily have $n = \mathrm{tr.\ deg.\ } \kappa(Q_1)/\kappa(P)$, where Q_1 is any minimal prime ideal in B not containing f. Hence in view of Thm. A.6 in the appendix we conclude $n = \dim B_Q + \mathrm{tr.\ deg.\ } \kappa(Q)/\kappa(P)$.)

Proof.

We delay the proof of the implication "(1) $\Rightarrow$ (2)" since this will follow from the proof of the less general Thm. 8.2.N which uses Cor. 7.1.1.N once we establish Cor. 7.1.1 in Chapter 9.

(2) $\Rightarrow$ (1). Let μ be as in (2). By Prop. 2.9 we have an exact sequence of homomorphisms of B_f-modules

(3) $\qquad \Gamma_A^1(C) \underset{C}{\otimes} B_f \to \Gamma_A^1(B_f) \to \Gamma_C^1(B_f) \to 0,$

where $C = A[T_1, \ldots, T_n]$. But $\Gamma_C^1(B_f) = 0$ by Prop. 4.3 in view of (2). Hence since $\{d(T_1), \ldots, d(T_n)\}$ generates the C-module $\Gamma_A^1(C)$, in view of (3) we conclude

(4) $\{d(\mu(T_1)), \ldots, d(\mu(T_n))\}$ generates the B_f-module $\Gamma_A^1(B_f)$.

In view of (2) we know that B_Q is a flat C_R-module, where $R = \mu^{-1}(QB_f)$. Hence

(5) B_Q is a flat A_P-module

since C_R is a flat A_P-module since C is a flat A-module. In view of (4) and (5), by Thm. 7.1 we obtain (1).

Finally we proceed to prove the remark following (2). Let f and μ be as in (2). From this point on we shall put instead $C = \kappa(P)[T_1, \ldots, T_n]$. Let Q_0 be the ideal in $B_f \otimes_A \kappa(P)$ generated by the image of Q. Put $\mu' = \mu \otimes_A \kappa(P) : C \to B_f \otimes_A \kappa(P)$ and put $R_0 = \mu'^{-1}(Q_0)$. We know

(6) $B_f \otimes_A \kappa(P)$ is étale over C at Q_0 via μ'

by (2) and Prop. 4.7. In particular,

(7) $(B_f \otimes_A \kappa(P))_{Q_0}$ is a flat C_{R_0}-module.

Since B_f is simple over A, without loss of generality we may replace Q by a minimal prime ideal in B not containing f. Hence

(8) Q_0 is a minimal prime ideal in $B_f \otimes_A \kappa(P)$.

Hence in view of (7) and (8), by Prop. A.8 in the appendix we conclude that R_0 is a minimal prime ideal in C. Hence by Cor. 7.5.2 and Thm. 7.5 we conclude

(9) $\quad\quad\quad\quad \kappa(R_0)$ is a separable field extension of $\kappa(P)$.

By (6) we know that $C' = ((B_f \otimes_A \kappa(P)) \otimes_C \kappa(R_0))_{Q_0'}$ is étale over $\kappa(R_0)$, where Q_0' is the ideal in $(B_f \otimes_A \kappa(P)) \otimes_C \kappa(R_0)$ generated by the image of Q_0. Hence by definition, $\Gamma^1_{\kappa(R_0)}(C') = 0$. Hence by Prop. 4.1 we conclude

(10) $\quad\quad \begin{cases} C' \text{ is a field and } C' \text{ is a finite} \\ \text{separable field extension of } \kappa(R_0). \end{cases}$

In view of (9) and (10) we conclude that C' is a separable field extension of $\kappa(P)$. Note $C' = \kappa(Q_0) = \kappa(Q)$ and $\kappa(R_0) = \kappa(R)$. Hence $\kappa(Q)$ is a separable field extension of $\kappa(P)$. Thus in view of (6), by Lemma 4.5 we conclude that $n = \text{tr. deg. } \kappa(Q)/\kappa(P)$. This establishes the remark.

8.2.N. **Theorem.** With notation and hypotheses as in Thm. 8.2, the implication "(2) ⇒ (1)" and the remark following (2) are true. Moreover, assume the added hypotheses of Cor. 7.1.1.N (for example, whenever either the ring A is noetherian or whenever there exists $g \in B$ such that $Q \in D(g)$ and B_g is a noetherian ring) the implication "(1) ⇒ (2)" is true.

Proof.

Since the implication "(2) ⇒ (1)" and the remark following (2) in Thm. 8.2 were just proved, to establish Thm. 8.2.N it remains only to establish the implication "(1) ⇒ (2)" under the added hypotheses of Cor. 7.1.1. N. First we assume that the ring A is reduced.

Let Q' be the ideal in $B \otimes_A \kappa(P)$ generated by the image of Q and let n be as in the remark following (2). By (1) and Cor. 7.1.1. N there exists $f_1 \in B$ such that $Q \in D(f_1)$ and such that B_{f_1} is simple over A. Hence by Cor. 7.5.1 we conclude that $\Gamma_A^1(B_{f_1})$ is a locally free B_{f_1}-module. Hence there exists $f_2 \in B$ such that $Q \in D(f_2)$ and such that $\Gamma_A^1(B_{f_2})$ is a free B_{f_2}-module. Put $f_3 = f_1 f_2$. Then we have that B_{f_3} is simple over A and $\Gamma_A^1(B_{f_3})$ is a free B_{f_3}-module. In view of (1), by definition we conclude that $(B \otimes_A \kappa(P))_{Q'}$ is simple over $\kappa(P)$. Hence by Lemma 7.3 we conclude that $\Gamma_{\kappa(P)}^1(B \otimes_A \kappa(P))_{Q'} = \Gamma_A^1(B_Q) \otimes_A \kappa(P)$ is a free $B_Q \otimes_A \kappa(P)$-module of rank n. Hence $\Gamma_A^1(B_{f_3})$ is a free B_{f_3}-module of rank n. Since $(B \otimes_A \kappa(P))_{Q'}$ is simple over $\kappa(P)$, by Thm. 7.1 there exists a subset $\{y_1, \ldots, y_n\}$ of $B_Q \otimes_A \kappa(P)$ such that

(3) $\quad \begin{cases} \{d(y_1), \ldots, d(y_n)\} \text{ is a basis for the free} \\ B_Q \otimes_A \kappa(P)\text{-module } \Gamma_A^1(B_Q) \otimes_A \kappa(P). \end{cases}$

Choose $x_i \in B_Q$ mapping to y_i in $B_Q \otimes_A \kappa(P)$ for each $1 \leq i \leq n$. In view of (3), by Nakayama's lemma we conclude that $\{d(x_1), \ldots, d(x_n)\}$ generates the B_Q-module $\Gamma_A^1(B_Q)$. For each $1 \leq i \leq n$ choose $b_i \in B$ and $s_i \in B \backslash Q$ such that $x_i = b_i / s_i$ and put $f_4 = f_3 s_1 \cdots s_n$. Then $\{x_1, \ldots, x_n\} \subset B_{f_4}$. Hence by Lemma A.14 in the appendix there exists $f_5 \in B_{f_4}$ such that $\{d(x_1), \ldots, d(x_n)\}$

generates the B_f-module $\Gamma_A^1(B_f)$, where $f = f_4 f_5$. Then B_f is simple over A, $\Gamma_A^1(B_f)$ is a free B_f-module of rank n, $\{x_1, \ldots, x_n\} \subset B_f$ and $\{d(x_1), \ldots, d(x_n)\}$ generates the free B_f-module $\Gamma_A^1(B_f)$.

Define a homomorphism of A-algebras $\mu : C \to B_f$, where $C = A[T_1, \ldots, T_n]$, by $\mu(T_i) = x_i$ for $1 \le i \le n$. Note B_f is a C-algebra of finite presentation under μ since B is a finitely presented A-algebra. (Actually we obtain the latter by replacing B by B_g for a suitable $g \in B$ such that $Q \in D(g)$.) By Prop. 2.9 we have an epimorphism $\Gamma_A^1(B_f) \to \Gamma_C^1(B_f)$ of B_f-modules. Hence since $\{d(x_1), \ldots, d(x_n)\}$ generates the B_f-module $\Gamma_A^1(B_f)$, we conclude that $\{d(x_1), \ldots, d(x_n)\}$ generates the B_f-module $\Gamma_C^1(B_f)$. But $d(x_i) = 0$ in $\Gamma_C^1(B_f)$ for $1 \le i \le n$ since $\mu(T_i) = x_i$ for $1 \le i \le n$. Hence $\Gamma_C^1(B_f) = 0$. Hence by Prop. 4.3 we conclude that B_f is unramified over C.

Hence in view of Prop. A.8 in the appendix, to establish (2) it suffices to show that

(4) $\qquad\qquad B_Q$ is a flat C_R-module,

where $R = \mu^{-1}(QB_f)$. Since B_f is simple over A at Q, by definition $B_Q \otimes_A \kappa(P)$ is simple over $\kappa(P)$. Hence by Lemma 8.1 we conclude that $B_f \otimes_A \kappa(P)$ is étale over $\kappa(P)[T_1, \ldots, T_n]$ at Q'. In particular, $B_Q \otimes_A \kappa(P)$ is a flat $\kappa(P)[T_1, \ldots, T_n]$-module. Also B_Q is a flat A_P-module by definition since B_f is simple over A. Hence by Thm. A.10 in the appendix we obtain (4). This establishes the implication "(1) $\Rightarrow$ (2)" for the case where the ring A is reduced.

Now, under the added hypotheses of Cor. 7.1.1.N, without assuming that the ring A is reduced, we proceed to establish the implication "(1) $\Rightarrow$ (2)". Let I be the nilradical of A. Since $\Gamma^1_{A/I}(B/IB) = \Gamma^1_A(B)/I \cdot \Gamma^1_A(B)$ by Prop. 2.11 and since Spec (A/I) (resp., Spec (B/IB)) identifies to Spec (A) (resp., Spec (B)) since I is the nilradical of A (resp., since IB is contained in the nilradical of B) by (1) and by definition we conclude that B/IB is simple over A/I at Q' where Q' denotes the image of Q in B/IB.

Since the ring A/I is reduced, by Thm. 8.2.N for the case where the ring A is reduced, which we proved above, we conclude that there exists $f_1 \in B$ such that $Q \in D(f_1)$, an integer $n \geq 0$ and an étale homomorphism $A/I[T_1, \ldots, T_n] \to B_{f_1}/IB_{f_1}$ of A/I-algebras. By Cor. 9.4.1 we conclude that there exists $f_2 \in B$ and an étale $A[T_1, \ldots, T_n]$-algebra C such that $C/IC = B_f/IB_f$, where $f = f_1 f_2$. (There is no vicious circle of proofs here since the proof of Cor. 9.4.1 relies only on Thm. 9.4 which in turn relies only on material through Chapter 7.) By Lemma A.21 in the appendix we conclude that C is isomorphic as $A[T_1, \ldots, T_n]$-algebra to B_f. This establishes the implication "(1) $\Rightarrow$ (2)" for the noetherian case and thus completes the proof of the theorem.

8.2.N.1. Remark. Here we proceed to establish the implication "(1) $\Rightarrow$ (2B)" in Cor. 7.5.1 under the additional assumption that the ring B is noetherian. Recall that we only established this implication in the special case where the ring A is reduced. The noetherian assumption on the ring B, of course, will be removed in 9.3 when we remove the noetherian assumption in Thm. 8.2.N.

Fix $Q \in \mathrm{Spec}(B)$. By (1) and Thm. 8.2.N we conclude that there exists $f \in B$ such that $Q \in D(f)$, an integer $n \geq 0$ and an étale homomorphism $\mu : A[T_1, \ldots, T_n] \to B_f$ of A-algebras. Since μ is an étale homomorphism of A-algebras, by Prop. 4.8 we conclude that $\Gamma_A^1(B)_f = \Gamma_A^1(A[T_1, \ldots, T_n]) \otimes_{A[T_1, \ldots, T_n]} B_f$. But by Prop. 2.8 $\Gamma_A^1(A[T_1, \ldots, T_n])$ is a free $A[T_1, \ldots, T_n]$-module whence $\Gamma_A^1(B)_f$ is a free B_f-module. This establishes (2B) in Cor. 7.5.1.

8.2.1. **Corollary.** Let A be a ring, B be an A-algebra with structure homomorphism $\alpha : A \to B$, C be a B-algebra with structure homomorphism $\beta : B \to C$, let $Q \in \mathrm{Spec}(C)$ and put $P = \beta^{-1}(Q)$. Suppose C is simple over B at Q and B is simple over A at P. Then C is simple over A at Q.

Proof.

By Thm. 8.2 there exists $g \in C$ such that $Q \in D(g)$, an integer $n \geq 0$ and a homomorphism $\nu : B' \to C_g$ of B-algebras such that C_g is étale over B' at QC_g, where $B' = B[T_1, \ldots, T_n]$, and there exists $f \in B$ such that $P \in D(f)$, an integer $m \geq 0$ and a homomorphism $\mu : A' \to B_f$ of A-algebras such that B_f is étale over A' at PB_f, where $A' = A[S_1, \ldots, S_m]$. Put $(B_f)' = B_f[T_1, \ldots, T_n]$, and $A'' = A'[T_1, \ldots, T_n]$.

Since C_g is an étale B'-algebra and $(B_f)' \otimes_{B'} C_g = (B')_f \otimes_{B'} C_g = C_{fg}$, by Cor. 4.7.1 we conclude

(1) $\begin{cases} \nu' = \nu \underset{B'}{\otimes} (B_f)' : (B_f)' \to C_{fg} \text{ is an étale} \\ \text{homomorphism of } B_f\text{-algebras.} \end{cases}$

Since B_f is an étale A'-algebra and $A'' \underset{A'}{\otimes} B_f = (B_f)'$, by Cor. 4.7.1 we conclude

(2) $\begin{cases} \mu' = \mu \underset{A'}{\otimes} A'' : A'' \to (B_f)' \text{ is an étale} \\ \text{homomorphism of A-algebras.} \end{cases}$

In view of (1), (2) and Cor. 4.6.1 we conclude that $\nu' \circ \mu' : A'' \to C_{fg}$ is an étale homomorphism of A-algebras whence by Thm. 8.2 we conclude that C is simple over A at Q. This establishes the corollary.

8.2.1.N. Corollary. With notation and hypotheses as in Cor. 8.2.1 we assume, in addition, that the ring B is noetherian. Then the conclusion of Cor. 8.2.1 holds.

Proof.

Same as for Cor. 8.2.1 except we use Thm. 8.2.N in place of Thm. 8.2.

8.2.1.1. Remark. At first sight it may seem pointless to state the less general Cor. 8.2.1.N above. However, the proof of Cor. 8.2.1 depends on Thm. 8.2 whose proof will not be completed until Chapter 9. Hence in many of the remaining corollaries in this chapter we make noetherian assumptions and add the suffix ".N" to the number in order to be able to refer to them later, knowing their proofs are, in fact, complete.

8.2.2. Corollary. Let A be a ring, B be an A-algebra and C be a B-algebra. Suppose C is simple over B and B is simple over A. Then C is simple over A.

Proof.
Immediate by Cor. 8.2.1.

8.2.2.N. Corollary. With notation and hypotheses as in Cor. 8.2.2 we assume, in addition, that the ring B is noetherian. Then the conclusion of Cor. 8.2.2 holds.

Proof.
Immediate by Cor. 8.2.1.N.

8.2.3. Corollary. Let A be a ring, B be an A-algebra with structure homomorphism $\alpha : A \to B$, let C be an A-algebra, put $\alpha' = \alpha \otimes_A C : C \to B \otimes_A C$, let $R \in \text{Spec}(B \otimes_A C)$ and put $Q = {\alpha'}^{-1}(R)$. Suppose B is simple over A at Q. Then $B \otimes_A C$ is simple over C at R.

Proof.
Since B is simple over A at Q, by Thm. 8.2 there exists $f \in B$ such that $Q \in D(f)$, an integer $n \geq 0$ and an étale homomorphism of A-algebras $\mu : A' \to B_f$, where $A' = A[T_1, \ldots, T_n]$. Put $C' = C[T_1, \ldots, T_n]$. Since μ is an étale homomorphism of A-algebras and $B_f \otimes_{A'} C' = B_f \otimes_A C$, by Cor. 4.7.1

we conclude that $\mu' = \mu \otimes_{A'} C' : C' \to B_f \otimes_A C$ is an étale homomorphism of C-algebras. Hence by Thm. 8.2 we conclude that $B \otimes_A C$ is simple over C at R.

8.2.3.1. **Remark.** With notation and hypotheses as in Cor. 8.2.3 let P denote the preimage of Q in A and suppose, in addition, that C_Q is a flat A_P-module. Then the converse of Cor. 8.2.3 holds, namely B is simple over A at Q if and only if $B \otimes_A C$ is simple over C at R. This appears later as Thm. 9.1 in Chapter 9.

8.2.3.N. **Corollary.** With notation and hypotheses as in Cor. 8.2.3 we assume, in addition, that the ring B is noetherian. Then the conclusion of Cor. 8.2.3 holds.

Proof.

Same as for Cor. 8.2.3 except we use Thm. 8.2.N in place of Thm. 8.2.

8.2.4. **Corollary.** Let A be a ring and B and C be A-algebras. Suppose B is simple over A. Then $B \otimes_A C$ is simple over C.

Proof.

Immediate by Cor. 8.2.3.

8.2.4.N. **Corollary.** With notation and hypotheses as in Cor. 8.2.4 we assume, in addition, that the ring B is noetherian. Then the conclusion of Cor. 8.2.4 holds.

Proof.

Immediate by Cor. 8.2.3.N.

8.2.5. Corollary. Let A be a ring and n be an integer ≥ 0. Then $A[T_1, \ldots, T_n]$ is a simple A-algebra.

Proof.

Immediate by Thm. 8.2.N.

CHAPTER 9

Descent of simple homomorphisms
and removal of all noetherian assumptions
in Chapter 7 and Chapter 8

In this chapter we show that the property of being a simple homomorphism descends well. (See Thm. 9.2 and its corollaries.) This enables us to prove Cor. 7.1.1 which succeeds in one stroke in removing all noetherian assumptions in chapters 7, 8 and 9 and completing the proofs of all preceding theorems, propositions and corollaries in chapters 7, 8 and 9. The fact that the property of being a simple homomorphism descends well is of great importance in algebraic geometry where the proof of theorems of great generality can be greatly simplified by first making a descent to the noetherian case. (Many classes of morphisms in algebraic geometry descend well.)

Also of interest in this chapter are Thm. 9.1 and Thm. 9.4, as well as Prop. 9.5 and Prop. 9.6 which were mentioned in Chapter 1. Witness the power of the Jacobian criterion in the proof of Thm. 9.4.

9.1. Theorem. Let A be a ring, B and C be A-algebras, let $R \in \text{Spec}(B \otimes_A C)$, let Q be the preimage of R in B, let R_0 be the preimage of R in C and let P be the preimage of Q in A. Suppose there exists $f \in B$ such that $Q \in D(f)$ and such that B_f is a finitely presented A-algebra. Suppose also that C_{R_0} is a flat A_P-module.

Then B is simple over A at Q if and only if $B \underset{A}{\otimes} C$ is simple over C at R.

Proof.

Assuming that $B \underset{A}{\otimes} C$ is simple over C at R it suffices to prove that B is simple over A at Q, since the converse is Cor. 8.2.3.

First we shall prove that B_Q is a flat A_P-module. Let $M \to N$ be a monomorphism of A_P-modules. Since C_{R_0} is a flat A_P-module we obtain a monomorphism $M \underset{A_P}{\otimes} C_{R_0} \to N \underset{A_P}{\otimes} C_{R_0}$ of C_{R_0}-modules. $(B \underset{A}{\otimes} C)_R = (B_Q \underset{A_P}{\otimes} C_{R_0})_R$ is a flat C_{R_0}-module since $B \underset{A}{\otimes} C$ is simple over C at R.
Hence we obtain a monomorphism

(1)
$$\begin{cases} M \underset{A_P}{\otimes} (B_Q \underset{A_P}{\otimes} C_{R_0})_R = \\ = (M \underset{A_P}{\otimes} C_{R_0}) \underset{C_{R_0}}{\otimes} (B_Q \underset{A_P}{\otimes} C_{R_0})_R \to \\ \to (N \underset{A_P}{\otimes} C_{R_0}) \underset{C_{R_0}}{\otimes} (B_Q \underset{A_P}{\otimes} C_{R_0})_R = \\ = N \underset{A_P}{\otimes} (B_Q \underset{A_P}{\otimes} C_{R_0})_R \end{cases}$$

of $(B \underset{A}{\otimes} C)_R$-modules. Since $(B_Q \underset{A_P}{\otimes} C_{R_0})_R$ is a faithfully flat B_Q-module since C_{R_0} is a flat A_P-module by hypothesis, we conclude in view of (1) that the homomorphism $M \underset{A_P}{\otimes} B_Q \to N \underset{A_P}{\otimes} B_Q$ of B_Q-modules is injective which proves

(2) $\qquad B_Q$ is a flat A_P-module.

Let Q' (resp., R') be the ideal generated by the image of Q (resp., R) in $B \otimes_A \kappa(P)$ (resp., $B \otimes_A \kappa(R_0)$). Now let k be a purely inseparable algebraic field extension of $\kappa(P)$ and let K be a compositum of k and $\kappa(R_0)$ in some field containing both k and $\kappa(R_0)$. Note that K is a purely inseparable algebraic field extension of $\kappa(R_0)$. Let Q'' (resp., R'') be the ideal generated by the image of Q' (resp., R') in B' (resp., B''), where $B' = (B \otimes_A \kappa(P))_{Q'} \otimes_{\kappa(P)} k$ and $B'' = (B \otimes_A \kappa(R_0))_{R'} \otimes_{\kappa(R_0)} K$; note

(3) $\quad \begin{cases} B' \text{ (resp., } B'' \text{) is a noetherian local ring with} \\ \text{maximal ideal } Q'' \text{ (resp., } R'' \text{) and the} \\ \text{homomorphism } B' \to B'' \text{ is local.} \end{cases}$

Since $(B \otimes_A C) \otimes_C \kappa(R_0) = B \otimes_A \kappa(R_0)$ and $B \otimes_A C$ is simple over C at R, by Lemma 7.7 we conclude

(4) $\qquad B''$ is a regular local ring.

Let Q_1 (resp., R_1) be the ideal generated by the image of Q' (resp., R') in $B \otimes_A k$ (resp., $B \otimes_A K$). Note that $B' = (B \otimes_A k)_{Q_1}$ and $B'' = (B \otimes_A K)_{R_1}$. $(B \otimes_A K)_{R_1}$ is a flat $(B \otimes_A k)_{Q_1}$-module since $B \otimes_A K$ is a flat $B \otimes_A k$-module. Hence

(5) $\qquad B''$ is a flat B'-module.

In view of (3), (4) and (5), by Lemma A.4 in the appendix we conclude

(6) $\qquad\qquad$ B' is a regular local ring.

In view of (2) and (6), by Lemma 7.7 we conclude that B is simple over A at Q. This completes the proof of the theorem.

9.1.N. Theorem. With notation and hypotheses as in Thm. 9.1 the implication "$\Leftarrow$" holds and if we assume, in addition, that the ring B is noetherian, then the implication "$\Rightarrow$" holds.

Proof.
Same as for Thm. 9.1 except we apply Cor. 8.2.3.N in place of Cor. 8.2.3 in order to establish the implication "$\Rightarrow$".

9.1.1. Corollary. Let A be a ring, B be a finitely presented A-algebra and let C be an A-algebra which is a fathfully flat A-module. Then B is simple over A if and only if $B \otimes_A C$ is simple over C.

Proof.
Immediate by Thm. 9.1.

9.1.1.N. Corollary. With notation and hypotheses as in Cor. 9.1.1 the implication "$\Leftarrow$" holds and if we assume, in addition, that the ring B is noetherian, then the implication "$\Rightarrow$" holds.

Proof.

Immediate by Thm. 9.1.N.

9.2. **Theorem.** Let A_0 be a ring, let $(A_i)_{i \in I}$ be a direct system of A_0-algebras indexed by a directed set I, fix $i_0 \in I$, let B_{i_0} and C_{i_0} be finitely presented A_{i_0}-algebras, for each $i \in I$ such that $i_0 \leq i$ put $B_i = B_{i_0} \otimes_{A_{i_0}} A_i$ and $C_i = C_{i_0} \otimes_{A_{i_0}} A_i$ yielding direct systems of A_{i_0}-algebras $(B_i)_{i \in I, i \geq i_0}$ and $(C_i)_{i \in I, i \geq i_0}$, put $A = \varinjlim_{i \in I} A_i$, $B = \varinjlim_{i \in I, i \geq i_0} B_i$ and $C = \varinjlim_{i \in I, i \geq i_0} C_i$ whence $B = B_{i_0} \otimes_{A_{i_0}} A$ and $C = C_{i_0} \otimes_{A_{i_0}} A$, let $\gamma_{i_0} : B_{i_0} \to C_{i_0}$ be a homomorphism of A_{i_0}-algebras, let $\gamma_i : B_i \to C_i$, $i \in I$, $i \geq i_0$ be the corresponding homomorphisms of A_i-algebras and let $\gamma : B \to C$ be the corresponding homomorphism of A-algebras. Then the following are true.

(1). Let $Q \in \text{Spec}(C)$ and for each $i \in I$, $i \geq i_0$ let Q_i be the preimage of Q in C_i. Then γ is simple at Q if and only if there exists $i \in I$, $i \geq i_0$ such that γ_i is simple at Q_i.

(2). γ is simple if and only if there exists $i \in I$, $i \geq i_0$ such that γ_i is simple.

Proof.

The implications "$\Leftarrow$" in (1) and (2) are immediate by Cor. 8.2.3 and Cor. 8.2.4 since $C = C_i \otimes_{B_i} B$ for each $i \in I$ such that $i \geq i_0$.

Next we proceed to establish the implication "$\Rightarrow$" in (1) (resp., (2)). By Thm. A.18 in the appendix we conclude by hypothesis that there exists

$i \in I$, $i \geq i_0$ such that $C_{i_{Q_i}}$ is a flat $B_{i_{P_i}}$-module where $P_i = \gamma_i^{-1}(Q_i)$ (resp., such that C_i is a flat B_i-module). Then by the proof of Thm. 9.1 we conclude that γ_i is simple at Q_i (resp., γ_i is simple). This completes the proof of the theorem.

9.2.N. **Theorem.** With notation and hypotheses as in Thm. 9.2 the implications "$\Rightarrow$" in (1) and (2) hold. Suppose, in addition, that there exists $i \in I$, $i \geq i_0$ such that B_i is a noetherian ring. Then the implications "$\Leftarrow$" in (1) and (2) hold.

Proof.

Same as for Thm. 9.2 except in place of Cor. 8.2.3 and Cor. 8.2.4 we use Cor. 8.2.3.N and Cor. 8.2.4.N.

9.2.1. **Corollary.** Let A_0 be a ring, A be an A_0-algebra, B be a finitely presented A-algebra and C be a finitely presented B-algebra with structure homomorphism $\gamma : B \to C$. Then the following are true.

(1). Then there exists a direct system of finitely presented A_0-algebras $(A_i)_{i \in I}$ (which can be taken to be subalgebras of A) indexed by a directed set I, $i_0 \in I$, a finitely presented A_{i_0}-algebra B_{i_0} and and a finitely presented B_{i_0}-algebra C_{i_0} with structure homomorph $\gamma_{i_0} : B_{i_0} \to C_{i_0}$ such that putting $B_i = B_{i_0} \otimes_{A_{i_0}} A$ and $C_i = C_{i_0} \otimes_{A_{i_0}} A$ for each $i \in I$, $i \geq i_0$ we obtain direct systems of A_{i_0}-algebras $(B_i)_{i \in I, i \geq i_0}$ and $(C_i)_{i \in I, i \geq i_0}$ such that putting $\gamma_i = \gamma_{i_0} \otimes_{A_{i_0}} A_i : B_i \to $ for each $i \in I$, $i \geq i_0$ the following are true:

(1A). $A = \varinjlim\limits_{i \in I} A_i$.

(1B). $B = \varinjlim\limits_{i \in I,\ i \geq i_0} B_i$.

(1C). $C = \varinjlim\limits_{i \in I,\ i \geq i_0} C_i$.

(1D). $\gamma = \varinjlim\limits_{i \in I,\ i \geq i_0} \gamma_i$.

(2). With notation as in (1) let $Q \in \mathrm{Spec}(C)$ and for each $i \in I$, $i \geq i_0$ let Q_i be the preimage of Q in C_i. Then γ is simple at Q if and only if there exists $i \in I$, $i \geq i_0$ such that γ_i is simple at Q_i.

(3). With notation as in (1), γ is simple if and only if there exists $i \in I$, $i \geq i_0$ such that γ_i is simple.

Proof.

Conclusion (1) is Thm. A.19 in the appendix. Hence (2) and (3) are immediate by (1) and Thm. 9.2.

9.2.1.N. Corollary. With notation and hypotheses as in Cor. 9.2.1, conclusion (1) of Cor. 9.2.1 holds and the implications "$\Rightarrow$" in (2) and (3) hold. Suppose, in addition, that there exists $i \in I$, $i \geq i_0$ such that B_i is a noetherian ring. Then the implications "$\Leftarrow$" in (2) and (3) hold.

Proof.

Same as for Cor. 9.2.1 except we use Thm. 9.2.N in place of Thm. 9.2.

9.2.2. Corollary. Let A_0 be a ring, B be an A_0-algebra and C be a finitely presented B-algebra with structure homomorphism $\gamma : B \to C$. Fix $Q \in \mathrm{Spec}(C)$. The following conditions are equivalent:

(1). γ is simple at Q.

(2). There exists a finitely presented sub-A_0-algebra B_0 of B, there exists a finitely presented B_0-algebra C_0 with structure homomorphism $\gamma_0 \in B_0 \to C_0$ such that $C = C_0 \otimes_{B_0} B$ and $\gamma = \gamma_0 \otimes_{B_0} B$ and such that letting Q_0 be the preimage of Q in C_0, γ_0 is simple at Q_0.

(3). Same as in (2) except we take $A_0 = \mathbb{Z}$.

Proof.

(1) $\Rightarrow$ (2). By Cor. 9.2.1.N we can take $B_0 = B_i$, $C_0 = C_i$ and $\gamma_0 = \gamma_i$ in conclusion (2) of Cor. 9.2.1.N.

(2) $\Rightarrow$ (3). Obvious.

(3) $\Rightarrow$ (1). Immediate by Cor. 8.2.3.N.

9.2.3. **Corollary.** Let A_0 be a ring, A be an A_0-algebra, B be a finitely presented A-algebra and C be a finitely presented B-algebra. The following conditions are equivalent:

(1). C is simple over B.

(2). There exists a finitely presented sub-A_0-algebra B_0 of B and a simple B_0-algebra C_0 such that $C = C_0 \otimes_{B_0} B$.

(3). Same as in (2) except we take $A_0 = \mathbb{Z}$.

Proof.

(1) $\Rightarrow$ (2). By Cor. 9.2.1.N we can take $B_0 = B_i$ and $C_0 = C_i$ in conclusion (3) of Cor. 9.2.1.N.

(2) ⇒ (3). Obvious.

(3) ⇒ (1). Immediate by Cor. 8.2.4.N.

9.3. Now we proceed to prove Cor. 7.1.1 which shall succeed in one stroke in removing all noetherian assumptions in chapters 7, 8 and 9 and completing the proofs of all preceding theorems, propositions and corollaries in chapters 7, 8 and 9.

Proof.

By hypothesis and Cor. 9.2.2 we conclude that there exist a subalgebra A_0 of A which is a finitely presented $\mathbb{Z}$-algebra and a finitely presented A_0-algebra B_0 such that $B = B_0 \otimes_{A_0} A$ and such that letting Q_0 be the preimage of Q in B_0, such that B_0 is simple over A_0 at Q_0. A_0, being a finitely presented $\mathbb{Z}$-algebra, is noetherian. Hence by Cor. 7.1.1.N we conclude that there exists $f_0 \in B_0$ such that $Q_0 \in D(f_0)$ and such that $B_{0_{f_0}}$ is simple over A_0. Let f denote the image of f_0 in B. Then $B_f = (B_0 \otimes_{A_0} A)_f = B_{0_{f_0}} \otimes_{A_0} A$. Hence by Cor. 9.2.3 we conclude that B_f is simple over A. Note $Q \in D(f)$. This completes the proof.

9.4. Theorem. Let A be a ring, I be an ideal in A, C be a simple A/I-algebra and let $P \in$ Spec (C). Then there exists $f \in C$ such that $P \in D(f)$ and such that there exists a simple A-algebra B such that C_f is isomorphic as an A/I-algebra to $B \otimes_A (A/I) = B/IB$.

Proof.

Since C is a finitely presented A/I-algebra there exist an integer $N \geq 1$ and a finitely generated ideal J in $(A/I)[T_1, \ldots, T_N]$ such that $C = A'/J$, where $A' = (A/I)[T_1, \ldots, T_N]$. Hence $P = Q/J$ for some $Q \in \mathrm{Spec}\,(A')$. Let $\{u_1, \ldots, u_r\}$ be a set of generators for the ideal J, let K denote the residue class field of A'_Q and let $\gamma : A' \to K$ be the canonical homomorphism. Since C is a simple A/I-algebra, and thus in particular simple over A/I at P, by the Jacobian criterion (condition (5) in Thm. 7.1) we conclude

(1) $$\mathrm{rank}_K (\gamma(\partial u_i/\partial T_j))_{1 \leq i \leq r,\ 1 \leq j \leq N} \geq N-n,$$

where $n = \sup_R \mathrm{tr.\,deg.}\ \kappa(R)/K$, where R runs through the set of minimal prime ideals in $C \otimes_{A/I} \kappa(Q_0)$, where Q_0 is the preimage of Q in A/I.

Put $A'' = A[T_1, \ldots, T_N]$ and let Q' be a prime ideal in A'' whose image in A' is Q such that Q lies over Q'. For each $1 \leq i \leq r$ choose $v_i \in A''$ mapping to u_i, put $J' = v_1 A'' + \ldots + v_r A''$, put $P' = Q'/J'$, put $C' = A''/J'$, let Q'_0 be the preimage of Q' in A, note that K is also the residue class field of $A''_{Q'}$, let $\gamma' : A'' \to K$ be the canonical homomorphism and put $n' = \sup_{R'} \mathrm{tr.\,deg.}\ \kappa(R')/K$, where R' runs through the set of minimal prime ideals in $C' \otimes_A \kappa(Q'_0)$. Note $n' \geq n$. By construction we conclude

$$\mathrm{rank}_K (\gamma'(\partial v_i/\partial T_j))_{1 \leq i \leq r,\ 1 \leq j \leq N} = \mathrm{rank}_K (\gamma(\partial u_i/\partial T_j))_{1 \leq i \leq r,\ 1 \leq j \leq N}$$

whence by (1) we conclude

(2) $$\mathrm{rank}_K (\gamma'(\partial v_i/\partial T_j))_{1 \leq i \leq r,\ 1 \leq j \leq N} \geq N-n.$$

But $N - n \geq N - n'$ since $n' \geq n$ whence by (2) we conclude

(3) $$\text{rank}_K(\gamma'(\partial v_i/\partial T_j))_{1\leq i\leq r,\ 1\leq j\leq N} \geq N - n'.$$

In view of (3), by the Jacobian criterion (condition (5) in Thm. 7.1) we conclude that C' is simple over A at P'. Hence by Cor. 7.1.1 there exists $f' \in C'$ such that $P' \in D(f')$ and such that $C'_{f'}$ is simple over A. Put $B = C'_{f'}$ and let f denote the image of f' in C. Evidently $C_f = B/IB$. This completes the proof of the theorem.

9.4.1. **Corollary.** With notation and hypotheses as in Thm. 9.4 except instead of assuming that C is a simple A/I-algebra we assume that C is an étale A/I-algebra, then the conclusion of Thm. 9.4 holds and, in addition, we can insist that B be an étale A-algebra.

Moreover, if, in addition, I is contained in the nilradical of A then having fixed f as in the statement of Thm. 9.4, B is uniquely determined as A-algebra up to isomorphism.

Proof.

By Thm. 9.4 there exist $f_0 \in C$ such that $P \in D(f_0)$ and a simple A-algebra B such that C_{f_0} is isomorphic as an A/I-algebra to $B \otimes_A (A/I)$. Let P' be the preimage of P in B, P_0 be the preimage of P in A/I and P'_0 be the preimage of P_0 in A. Since C is étale over A/I at P and since the fibers $\kappa(P'_0) \to B \otimes_A \kappa(P'_0)$ and $\kappa(P_0) \to C_f \otimes_{A/I} \kappa(P_0)$ of the homomorphisms $A \to B$ and $A/I \to C_f$ containing P are thus isomorphic we conclude by definition that B is unramified over A at P' and therefore by Prop. 4.4 also in an affine open neighborhood D(g) of P' in B, g ∈ B. Hence B_g is étale over A. Choose $n \geq 1$ and c ∈ C such that c/f^n is the image of g in C_f. Hence taking $f = f_0 c$ suffices. In view of Lemma A.21 in the appendix this establishes the corollary.

9.5. **Proposition.** Let $m \geq 1$, $N \geq 1$, $\{f_1, \ldots, f_m\}$ be a subset of $\mathbb{Z}[T_1, \ldots, T_N]$, let p ∈ $\mathbb{Z}$ be a prime number, F_p be the finite field with p elements, k be a field of characteristic 0 and F be a field of characteristic p. Then the following are true.

(1). $k[T_1, \ldots, T_N]/(f_1, \ldots, f_m)$ is simple over k if and only if $\mathbb{Q}[T_1, \ldots, T_N]/(f_1, \ldots, f_m)$ is simple over $\mathbb{Q}$.

(2). $F[T_1, \ldots, T_N]/(f_1, \ldots, f_m)$ is simple over F if and only if $F_p[T_1, \ldots, T_N]/(f_1, \ldots, f_m)$ is simple over F_p.

Proof.

Immediate by Cor. 9.1.1.

9.6. **Proposition.** Let $m \geq 1$, $N \geq 1$ and $\{f_1, \ldots, f_m\}$ be a subset of $\mathbb{Z}[T_1, \ldots, T_N]$. Then the following conditions are equivalent:

(1). $\mathbb{Q}[T_1, \ldots, T_N]/(f_1, \ldots, f_m)$ is simple over $\mathbb{Q}$.

(2). $F_p[T_1, \ldots, T_N]/(f_1, \ldots, f_m)$ is simple over F_p for all but finitely many prime numbers p.

(3). $F_p[T_1, \ldots, T_N]/(f_1, \ldots, f_m)$ is simple over F_p for infinitely many prime numbers p.

Proof.

Put $A = \mathbb{Z}[T_1, \ldots, T_N]/(f_1, \ldots, f_m)$ and $B = \mathbb{Q}[T_1, \ldots, T_N]/(f_1, \ldots, f_m)$. Note that Spec (B) identifies to the generic fiber of Spec (A) over $\mathbb{Z}$, that is the set of prime ideals in A which contain no prime number.

(1) $\Rightarrow$ (2). Since by hypothesis A is simple over $\mathbb{Z}$ at all points of the generic fiber, by Cor. 7.1.1 there exists an open neighborhood U of the generic fiber in Spec (A) such that A is simple over $\mathbb{Z}$ at all points of U. Choose U to be the largest such open set. Choose a set I and $g_i \in A$ for each $i \in I$ such that $U = \bigcup_{i \in I} D(g_i)$. Put $J = \Sigma_{i \in I} g_i A$. Hence given $Q \in$ Spec (A), denoting the image of Q in A/J by Q' we have $Q \notin U$ if and only if $Q' \in$ Spec (A/J) and there exists a prime number $p_{Q'}$ such that $p_{Q'} \in Q'$. But A/J, being noetherian, has only finitely many minimal prime ideals. Hence the set $\{p_{Q'} | Q \in$ Spec (A)$\setminus U\}$ is finite. This establishes (2).

(2) $\Rightarrow$ (3). Obvious.

(3) $\Rightarrow$ (1). Let Q be a maximal ideal in B and P be the preimage of Q in A. To establish (1), by Cor. 7.1.1 it suffices to show that B is simple over $\mathbb{Q}$ at Q. But by definition it suffices to show that A is simple over $\mathbb{Z}$ at P. By Thm. A.7 in the appendix there exists a nonempty open subset U of Spec (A/P) such that $\{(p) | p \in \mathbb{Z}$ is a prime number and $p \in P_1$ for some $P_1 \in U\}$ is open in Spec ($\mathbb{Z}$) and thus contains all but finitely many prime ideals in $\mathbb{Z}$. Since also the preimage of no maximal ideal of the generic fiber B of A over $\mathbb{Z}$ in A is a maximal ideal of A, in view of (3) we conclude that there exist a prime number p and $P_1 \in$ Spec (A) such that $P \subset P_1$, such that

$p \in P_1$ and such that $F_p[T_1, \ldots, T_N]/(f_1, \ldots, f_m)$ is simple over F_p. Hence by definition A is simple over $\mathbb{Z}$ at P_1. Hence by Cor. 7.1.1 we conclude that A is simple over $\mathbb{Z}$ at P. This establishes (1).

CHAPTER 10

Simple morphisms of preschemes and translation of previous theorems into the language of preschemes

In this chapter are the most important theorems of the preceding chapters translated into the language of preschemes. Since the proofs, of course, reduce immediately to the affine case, no further proof is required for the versions presented in this chapter. We begin with the preliminaries of defining simple, unramified and étale morphisms of preschemes. It is assumed here that the reader is already well-acquainted with the language and fundamentals of preschemes.

10.1. Definition. Let k be a field, X be a prescheme over k and $x \in X$. We say that X is simple over k at x if and only if there exists an affine open neighborhood U of x in X such that the ring of U is simple over k at the prime ideal corresponding to x. We say that X is simple over k if and only if X is simple over k at each point x of X.

10.2. Definition. Let X and Y be preschemes, $f : X \to Y$ be a morphism of preschemes, $x \in X$ and put $y = f(x)$. We say that f is simple at x or X is simple over Y at x if and only if f is locally of finite presentation in an open neighborhood of x in X, f is flat at x and $X \times_Y \text{Spec}(\kappa(y))$ is simple over $\kappa(y)$ at x. We say that f is simple or X is simple over Y if and only if f is simple at each point x of X.

10.3. **Definition.** Let X, Y, f & x be as in Def. 10.2. We say that **f is unramified at x** or **X is unramified over Y at x** if and only if f is locally of finite presentation in an open neighborhood of x in X and $\Gamma_Y^1(X)_x = 0$. (See Remark 10.3.1 below for the definition of the sheaf of $\mathcal{O}_X$-modules $\Gamma_Y^1(X)$.) We say that **f is étale at x** or **X is étale over Y at x** if and only if f is flat and unramified at x. We say that **f is unramified** or **X is unramified over Y** (resp., **f is étale** or **X is étale over Y**) if and only if f is unramified (resp., étale) at all points x of X.

10.3.1. **Remark.** Let X, Y and f be as in Def. 10.2. The diagonal morphism $X \to X \underset{Y}{\times} X$ being an immersion is a closed immersion $X \to V$ for some open subset V of $X \underset{Y}{\times} X$. Let I be the sheaf of ideals of $\mathcal{O}_X$ defining the closed subprescheme of X corresponding to the diagonal immersion into V. The sheaf of $\mathcal{O}_X$-modules I/I^2 is denoted $\Gamma_Y^1(X)$ and is called the **sheaf of Kähler 1-differentials** of the Y-prescheme X.

Note that if X = Spec (B) and Y = Spec (A) are affine then by construction and Def. 2.1 we conclude that $\Gamma_Y^1(X) = \Gamma_A^1(B)^\sim$. Moreover, in the general case given $x \in X$, putting $y = f(x)$, choosing an affine open neighborhood W of y in Y with ring A and an affine open neighborhood U of x in X with ring B and letting $Q \in $ Spec (B) be the prime ideal corresponding to x we have $\Gamma_Y^1(X)_x = \Gamma_A^1(B)_Q$. Hence if X and Y are both affine then Def. 10.1, Def. 10.2 and Def. 10.3 agree with the corresponding definitions in earlier chapters.

Note that in all cases $\Gamma_Y^1(X)$ is a quasicoherent sheaf of $\mathcal{O}_X$-modules and that in the case where f is locally of finite presentation, then by Cor. 2.10.2 we conclude that $\Gamma_Y^1(X)$ is a finitely presented sheaf of $\mathcal{O}_X$-modules.

10.3.2. **Remark.** Let X, Y and f be as in Def. 10.2. Let $p_i : X \times_Y X \to X$, $i = 1, 2$ be the projection morphisms and for $i = 1, 2$ let $\varphi_i : \mathcal{O}_X \to p_{i_*}(\mathcal{O}_{X \times_Y X})$ be the homomorphism of sheaves of $\mathcal{O}_X$-modules defined by p_i. Since $p_{1_*}(\mathcal{O}_{X \times_Y X})$ and $p_{2_*}(\mathcal{O}_{X \times_Y X})$ are canonically isomorphic sheaves of $\mathcal{O}_X$-modules we can identify them and thus define a homomorphism $\varphi = \varphi_1 - \varphi_2$ of sheaves of abelian groups on X. By construction we have that Im(φ) is a subsheaf of abelian groups of I. Let $\varphi' : \mathcal{O}_X \to I$ be the homomorphism of sheaves of abelian groups defined by φ. Put

$$d = d_{X/Y} = (\mathcal{O}_X \xrightarrow{\varphi'} I \to I/I^2 = \Gamma_Y^1(X)).$$

If X = Spec (B) and Y = Spec (A) are affine we have $d_{B/A}^{\sim} = d_{X/Y} : B^{\sim} \to \Gamma_A^1(B)^{\sim}$ is the homomorphism of sheaves of abelian groups defined by the homomorphism of A-modules $d_{B/A} : B \to \Gamma_A^1(B)$ defined in Def. 2.1. Moreover, in the general case given $x \in X$, putting $y = f(x)$, choosing an affine open neighborhood W of y in Y with ring A and an affine open neighborhood U of x in X with ring B and letting $Q \in$ Spec (B) be the prime ideal corresponding to x we have that $(d_{X/Y})_x : B_Q \to \Gamma_A^1(B)_Q$ is the homomorphism obtained from $d_{B/A}$ by passing to quotients.

10.4. **Proposition.**

   (1). Let $f : X \to Y$ and $g : Y \to Z$ be morphisms of preschemes, let $x \in X$ and put $y = f(x)$. Suppose f is simple at x and g is simple at y. Then $g \cdot f$ is simple at x.

   (2). Let $f : X \to Y$ and $g : Y' \to Y$ be morphisms of preschemes, put $X' = X \times_Y Y'$, $f' = f \times_Y Y'$ and $g' = g \times_Y X$, let $x' \in X'$ and put $x = g'(x')$. Suppose f is simple at x. Then f' is simple at x'.

Proof.

Since (1) and (2) are local on X and Y we may assume that both X and Y are affine whence (1) is immediate by Cor. 8.2.1 and (2) is immediate by Cor. 8.2.3.

10.4.1. **Corollary.**

   (1). Let $f : X \to Y$ and $g : Y \to Z$ be morphisms of preschemes. Suppose both f and g are simple. Then $g \cdot f$ is simple.

   (2). Let $f : X \to Y$ and $g : Y' \to Y$ be morphisms of preschemes and put $f' = f \times_Y Y'$. Suppose f is simple. Then f' is simple.

Proof.

Immediate by Prop. 10.4.

10.5. **Theorem.** Let X and Y be preschemes and $f : X \to Y$ be a morphism of preschemes. Then the following two conditions are equivalent:

(1). f is simple.

(2). The following four conditions hold:

   (2A). f is locally of finite presentation.

   (2B). $\Gamma^1_Y(X)$ is a locally free sheaf of $\mathcal{O}_X$-modules.

   (2C). For each $x \in X$, putting $y = f(x)$ and each generic point z of the fiber $X \underset{Y}{\times} \operatorname{Spec}(\kappa(y)) = f^{-1}(y)$ we have that $\kappa(z)$ is a separable field extension of $\kappa(y)$.

   (2D). f is flat.

Proof.

Since all four conditions in (2) as well as condition (1) are local on both X and Y we may assume without loss of generality that both X and Y are affine whence the theorem is immediate by Cor. 7.5.1.

10.6. Theorem. Let X and Y be preschemes, $f : X \to Y$ be a morphism of preschemes, $x \in X$ and put $y = f(x)$. The following two conditions are equivalent:

   (1). f is simple at x.

   (2). There exists an open neighborhood U of x in X, an integer $n \geq 1$ and an étale morphism $g : U \to Y[T_1, \ldots, T_n]$ of Y-preschemes.

Moreover, in condition (2) we necessarily have $n = \operatorname{tr. deg.} \kappa(z)/\kappa(y)$ where z is any generic point of any affine open neighborhood of x contained in U. Moreover, in condition (2) n can be characterized by $n = \dim(\mathcal{O}_{X,x}) + \operatorname{tr. deg.} \kappa(x)/\kappa(y)$.

Proof.

Since both (1) and (2) are local on both X and Y we may assume without loss of generality that both X and Y are affine whence the theorem is immediate by Thm. 8.2.

10.6.1. Corollary. Let X and Y be preschemes, $f : X \to Y$ be a morphism of preschemes, and let $x \in X$. Suppose f is simple at x. Then there exists an open neighborhood U of x in X such that f is simple at all points of U.

Proof.

Immediate by Thm. 10.6 since we may choose U as in condition (2) of Thm. 10.6.

10.6.2. Corollary. Let X be a prescheme and n be an integer ≥ 0. Then $X[T_1, \ldots, T_n]$ is simple over X.

Proof.

Immediate by condition (2) of Thm. 10.6.

10.7. Theorem. Let k be a field, X be a prescheme locally of finite type over k, let $x \in X$ and put n = sup tr. deg. $\kappa(z)/k$ where z runs through the set of generic points of $\text{Spec}(\mathcal{O}_{X,x})$. (Note n = $\dim(\mathcal{O}_{X,x})$ + tr. deg. $\kappa(x)/k$) The following conditions are equivalent:

(1). X is simple over k at x.

(2). The minimum cardinality of a set of generators for $\Gamma_k^1(X)_x$ as an $\mathcal{O}_{X,x}$-module is $\leq n$.

(3). The minimum cardinality of a set of generators for $\Gamma_k^1(X)_x$ as an $\mathcal{O}_{X,x}$-module equals n.

(4). $\Gamma_k^1(X)_x$ is a free $\mathcal{O}_{X,x}$-module of rank n.

(5). There exists a subset $\{a_1, \ldots, a_n\}$ of $\mathcal{O}_{X,x}$ which is algebraically independent over k such that $\Gamma_k^1(X)_x$ is a free $\mathcal{O}_{X,x}$-module of rank n with basis $\{d(a_1), \ldots, d(a_n)\}$.

Proof.

Since all conditions are local in an open neighborhood of x in X, without loss of generality we may replace X by an affine open neighborhood of x in X whose ring is a finitely generated k-algebra. The theorem is now immediate by definition and by Thm. 7.1.

10.8. Theorem. (The Jacobian criterion) Let k be a field, X be a prescheme locally of finite type over k and let $x \in X$. Choose an affine open neighborhood U of x in X such that the ring B of U is a finitely generated k-algebra. Hence there exist an integer $N \geq 1$ and a finitely generated ideal I of $k[T_1, \ldots, T_N]$ such that $B = k[T_1, \ldots, T_N]/I$. Let $M \geq 1$ and $\{f_1, \ldots, f_M\}$ be a set of generators for the ideal I in $k[T_1, \ldots, T_N]$. Let $\mathfrak{a}$ denote the Jacobian matrix $(\partial f_i / \partial T_j)_{1 \leq i \leq M, 1 \leq j \leq N}$ and denote by $\mathfrak{a}(x)$ the matrix obtained from $\mathfrak{a}$ by taking the image of each entry of $\mathfrak{a}$ in $\kappa(x)$ under the canonical homomorphism $B \to \kappa(x)$.

Put $n = \sup \text{tr. deg. } \kappa(z)/k$ where z runs through the set of generic points of Spec $(\mathcal{O}_{X,x})$. (Note $n = \dim(\mathcal{O}_{X,x}) + \text{tr. deg. } \kappa(x)/k$.) Then the following conditions are equivalent:

(1). X is simple over k at x.

(2). $\text{rank}_{\kappa(x)} \mathfrak{a}(x) \geq N - n$.

(3). $\text{rank}_{\kappa(x)} \mathfrak{a}(x) = N - n$.

Proof.

Since all conditions are local in an open neighborhood of x in X, without loss of generality we may replace X by Spec (B). The theorem is now immediate by Thm. 7.1.

10.9. Theorem. Let k be a field, X be a prescheme locally of finite type over k and let $x \in X$. The following conditions are equivalent:

(1). X is simple over k at x.

(2). For each purely inseparable algebraic field extension k' of k, the local ring $\mathcal{O}_{X,x} \otimes_k k'$ is regular.

(3). For each finite purely inseparable algebraic field extension k' of k, the local ring $\mathcal{O}_{X,x} \otimes_k k'$ is regular.

(4). For each extension $k' = k^{p^{-s}}$ of k where p is the characteristic exponent of k and s is an integer > 0, the local ring $\mathcal{O}_{X,x} \otimes_k k'$ is regular.

(5). The local ring $\mathcal{O}_{X,x} \otimes_k k^{p^{-\infty}}$ is regular, where p is the characteri exponent of k.

(6). For each perfect field extension k' of k, all the local rings of $\mathcal{O}_{X,x} \otimes_k k'$ are regular.

(7). For each finite field extension k' of k, all the local rings of $\mathcal{O}_{X,x} \otimes_k k'$ are regular.

(8). For each field extension k' of k such that k' is a finitely generated k-algebra, all the local rings of $\mathcal{O}_{X,x} \otimes_k k'$ are regular.

(9). For each field extension k' of k, all the local rings of $\mathcal{O}_{X,x} \otimes_k k'$ are regular.

Proof.

By replacing X by an affine open neighborhood of x whose ring is a finitely generated k-algebra, we obtain the theorem as an immediate consequence of Thm. 7.8.

10.10. **Theorem.** Let Y be a prescheme, Y_0 be a closed subprescheme of Y, X_0 be a simple Y_0-prescheme and let $x_0 \in X_0$. Then there exists an affine open neighborhood U_0 of x_0 in X_0 and there exists a simple affine Y-prescheme U such that U_0 is isomorphic to $U \times_Y Y_0$ as Y_0-preschemes.

Proof.

Let y_0 be the image of x_0 in Y_0, V be an affine open neighborhood of y_0 in Y, put $V_0 = V \cap Y_0$ and let W_0 be an affine open neighborhood of x_0 in X_0. Let A be the ring of V and C be the ring of W_0. Since V_0 is a closed subprescheme of the affine scheme V with ring A, the ring of V_0 is A/I where I is an ideal in A.

Hence by Thm. 9.4 we conclude that there exist $f \in C$ such that $x_0 \in \operatorname{Spec}(C_f)$ and a simple A-algebra B such that C_f is isomorphic as A/I-algebra to $B \otimes_A A/I$. Put $U_0 = \operatorname{Spec}(C_f)$ and $V = \operatorname{Spec}(B)$. Then we have $U_0 = U \times_V V_0$ and $V_0 = V \times_Y Y_0$ whence $U_0 = U \times_Y Y_0$ by transitivity of fiber products. This completes the proof of the theorem.

10.11. **Theorem.** Let $f : X \to Y$ be a map of preschemes locally of finite presentation, $g : Y' \to Y$ be a map of preschemes, put $X' = X \times_Y Y'$, $f' = f \times_Y Y'$ and $g' = g \times_Y X$. Let $x' \in X'$, put $x = g'(x')$, and $y' = f'(x')$. Suppose g is flat at y'. Then f is simple at x if and only if f' is simple at x'.

Proof.

Since the conditions are local on X, Y and Y' we may assume without loss of generality that X, Y and Y' are affine whence the theorem is immediate by Thm. 9.1.

10.11.1. **Corollary.** Let $f : X \to Y$ be a map of preschemes locally of finite presentation, $g : Y' \to Y$ be a map of preschemes, put $X' = X \times_Y Y'$ and $f' = f \times_Y Y'$. Suppose g is faithfully flat, that is flat and surjective. Then f is simple if and only if f' is simple.

Proof.

Immediate by Thm. 10.11.

10.12. **Theorem.** Let A be a ring, Y be an A-prescheme and X be a Y-prescheme locally of finite presentation. Fix $x \in X$ and let y denote the image of x in Y. Then the following conditions are equivalent:

(1). X is simple over Y at x.

(2). There exists an affine open neighborhood V of y in Y and an affine open neighborhood U of x in X contained in the preimage of V in X and there exists an affine A-prescheme Y_0 of finite presentation such that V is a Y_0-prescheme and there exists a simple affine Y_0-prescheme X_0 such that $U = X_0 \times_{Y_0} V$.

(3). Same as in (2) except we take $A = \mathbb{Z}$.

Proof.

Since all conditions are local we may assume without loss of generality that X and Y are affine, and thus in (1) in view of Cor. 10.6.1 we may assume that X is simple over Y. Hence the theorem is an immediate consequence of Cor. 9.2.3.

APPENDIX

A.1. Theorem. Let A be a complete local ring containing a field with residue class field K. The following are true:

(1). A contains a field L mapped onto K by the canonical homomorphism $A \to K$.

(2). If, in addition, K is a finitely generated separable field extension of k we can choose L in (1) so that it contains the subfield k of A.

Proof.

For a proof see [4], p. 205.

A.2. Lemma. Let A be a unique factorization domain. Then all minimal prime ideals of height 1 in A are principal and are the ones generated by irreducible elements.

Proof.

Let $P \in \text{Spec}(A)$ be of height 1. Since $P \neq (0)$ and $P \neq A$ since P is prime, P must contain a nonunit a. Let $a = a_1 \cdots a_n$ be a factorization of a into irreducible elements of A. Since $a_1 \cdots a_n = a \in P$ and P is a prime ideal we conclude that one of the a_i, $1 \leq i \leq n$, say a_1 is an element of P. Hence $Aa_1 \subset P$. $(0) \neq Aa_1$ since $a_1 \neq 0$, being irreducible. Aa_1 is a prime ideal since A is a unique factorization domain. Hence $P = Aa_1$ since P is of height 1. This establishes the lemma.

A.3. Theorem. Let A be a noetherian local ring of dimension r with maximal ideal $\mathcal{m}$. Then r is the smallest nonnegative integer s such that there exists a subset $\{x_1, \ldots, x_s\}$ of $\mathcal{m}$ and an integer $N \geq 1$ such that $\mathcal{m}^N \subset Ax_1 + \ldots + Ax_s$.

Proof.

For a proof see [5].

A.3.1. Corollary. Let A be a noetherian local ring of dimension r with maximal ideal $\mathcal{m}$ and put $K = A/\mathcal{m}$. Then $\mathcal{m}/\mathcal{m}^2$ is a K-vector space of dimension $\geq r$.

Proof.

By Nakayama's lemma $\dim_K \mathcal{m}/\mathcal{m}^2$ is the minimum cardinality n of a set of generators for $\mathcal{m}$ as an ideal in A and by Thm. A.3 we conclude that $n \geq r$.

A.4. Lemma. Let A and A' be noetherian local rings and $A \to A'$ be a local homomorphism of local rings. Suppose A' is a flat A-module and A' is a regular local ring. Then A is a regular local ring.

Proof.

For a proof see Prop. 17.3.3 in Chap. 0 in EGA IV, [2] in the bibliography.

A.5.1. Lemma. Let k be a field, n be an integer ≥ 1 and P, Q $\in$ Spec($k[T_1, \ldots, T_n]$) such that $P \subsetneq Q$ and such that P' $\in$ Spec($k[T_1, \ldots, T_n]$)

and $P \subset P' \subset Q$ implies that $P' = P$ or $P' = Q$. Then tr. deg. $\kappa(P)/k \geq$ tr. deg. $\kappa(Q)/k + 1$.

Proof.

Put $A = k[T_1, \ldots, T_n]/P$, let K be the quotient field of A and L be the residue class field of A_Q. Note that $\kappa(P) = K$ and $\kappa(Q) = L$. Put $r = $ tr. deg. L/k. Let $\{x_1, \ldots, x_r\}$ be a transcendence basis for L over k. For each $1 \leq i \leq r$ choose $a_i \in A$ and $s_i \in A \setminus Q$ such that $y_i = a_i/s_i$ maps to x_i under the canonical homomorphism $A_Q \to L$. Then $\{y_1, \ldots, y_r\}$ is algebraically independent over k and more strongly, putting $Q' = QA_Q$,

(1) $\left\{ \begin{array}{l} \text{If a polynomial in } y_1, \ldots, y_r \text{ with coefficients} \\ \text{in } k \text{ is an element of } Q', \text{ then all coefficients} \\ \text{of this polynomial are zero.} \end{array} \right.$

In view of (1) we conclude that tr. deg. $K/k \geq r$.

Fix $0 \neq t \in Q$. We proceed to show that t is transcendental over F, where $F = k(y_1, \ldots, y_r)$. This will show that tr. deg. $K/k \geq r + 1$ and complete the proof of the lemma. Indeed, suppose, on the contrary, that t is algebraic over F. Let $T^n + f_{n-1}T^{n-1} + \ldots + f_1 T + f_0$ be the minimal polynomial for t over F. Hence evaluating at t yields $t^n + f_{n-1}t^{n-1} + \ldots + f_1 t + f_0 = 0$ in K. By clearing denominators in the f_i, $0 \leq i \leq n-1$, we obtain $g_i \in k[y_1, \ldots, y_r]$ for $0 \leq i \leq n$ such that $g_n t^n + g_{n-1} t^{n-1} + \ldots + g_1 t + g_0 = 0$ in A_Q. Solving for g_0 yields $g_0 \in Q'$ since $t \in Q'$. In view of (1) we conclude that $f_0 = 0$. Hence $t(g_n t^{n-1} + g_{n-1} t^{n-2} + \ldots + g_1) = 0$ in A_Q. Since $t \neq 0$ we conclude that $g_n t^{n-1} + g_{n-1} t^{n-2} + \ldots + g_1 = 0$ in A_Q. Repeating this argument yields $g_1 = 0$ and iterating the process yields $g_0 = g_1 = \ldots = g_n = 0$. Hence $f_0 = f_1 = \ldots = f_{n-1} = 0$. Hence $t^n = 0$ in K implying $t = 0$, a contradiction. Hence t is indeed transcendental over F. This completes the proof of the lemma.

A.5.2. Lemma. With notation and hypotheses as in Lemma A.5.1 we conclude that tr. deg. $\kappa(P)/k$ = tr. deg. $\kappa(Q)/k + 1$.

Proof.

Let A, K and L be as in the proof of Lemma A.5.1, put r = tr. deg. K/k and for each $1 \leq i \leq n$ let x_i be the image of T_i in A.

Case 1. $r = n$.
Hence $\{x_1, \ldots, x_n\}$ is algebraically independent over k and A is a polynomial ring in n variables. Since A is a unique factorization domain and $Q \in \text{Spec}(A)$, by Lemma A.2 we conclude that $Q = Af$ for some irreducible element f of A. deg $f \neq 0$ since $Q \neq A$. Hence at least one of the x_i, $1 \leq i \leq n$ occurs in the formal polynomial expression of f, say x_n. Then Q contains no polynomial independent of x_n since $Q = Af$. Hence the Q residues of $x_1, \ldots, x_{n-1}$ are algebraically independent over k. Hence tr. deg. $L/k \geq n - 1$. But tr. deg. $L/k \leq n - 1$ by Lemma A.5.1. This completes the proof of the lemma for case 1.

Case 2. $r < n$.

By the normalization lemma we can choose r elements $y_1, \ldots, y_r$ in A such that A is integral over $B = k[y_1, \ldots, y_r]$. Put $Q_0 = Q \cap B$. Then B is a polynomial ring in r variables. Since B is integrally closed, being a unique factorization domain, and Q is a prime ideal of height 1 in A by hypothesis, Q_0 is necessarily of height 1 in B and hence by Case 1 we have tr. deg. $(B/Q_0)_{(0)}/k = r - 1$. But A/Q is integral over B/Q_0 since A is integral over B. Put $S = B/Q_0 \setminus \{0\}$. Hence $S^{-1}(A/Q)$ is integral over $(B/Q_0)_{(0)}$ whence tr. deg. $L/k \geq$ tr. deg. $(B/Q_0)_{(0)}/k = r - 1$. (Note $L = (S^{-1}(A/Q))_{(0)}$.) But tr. deg. $L/k \leq r - 1$ by Lemma A.5.1. This establishes Case 2 and completes the proof of the lemma.

A.6. Theorem. Let k be a field, B be a finitely generated k-algebra, $Q \in \operatorname{Spec}(B)$ and put $A = B_Q$. Then the following are true.

(1). $\dim B = \sup \dim(B_\mathfrak{m})$, where $\mathfrak{m}$ runs through the set of maximal ideals of B.

(2). $\dim B = \sup$ tr. deg. $\kappa(P)/k$, where P runs through the set of minimal prime ideals of B.

(3). $\dim A +$ tr. deg. $\kappa(A)/k = \sup$ tr. deg. $\kappa(P)/k$, where P runs through the set of minimal prime ideals of B contained in Q, that is the set of minimal prime ideals of A.

(4). Let P_0 be a minimal prime ideal of A. Then $\dim A = \dim(A/P_0)$ if and only if tr. deg. $\kappa(P_0)/k = \sup$ tr. deg. $\kappa(P)/k$, where P runs through the set of minimal prime ideals of A.

Proof.

Note that (1) is immediate by definition. First we proceed to show that (3) implies (2). Let $\mathcal{m}$ be a maximal ideal of B. $B/\mathcal{m}$ is a finitely generated k-algebra since B is a finitely generated k-algebra. Hence $B/\mathcal{m}$ is algebraic over k, that is tr. deg. $\kappa(\mathcal{m})/k = 0$. Hence by (3) we conclude

(5) $\begin{cases} \dim B_{\mathcal{m}} = \sup \text{tr. deg. } \kappa(P)/k, \text{ where } P \text{ runs} \\ \text{through the set of minimal prime ideals of } B \\ \text{contained in } \mathcal{m}. \end{cases}$

In view of (1) and (5) we obtain (2).

Next we proceed to show that (3) for the case where A is an integral domain implies (4). Given $P_0 \in \text{Spec}(A)$, $A/P_0 = B_Q/P_0 = (B/P_0)_Q$, with abuse of notation, whence A/P_0 is the localization at a prime ideal of an integral domain which is a finitely generated k-algebra. Hence (3) for the case where A is an integral domain applied to A/P_0 yields $\dim(A/P_0) + $ tr. deg. $\kappa(A/P_0)/k = $ tr. deg. $((A/P_0)_{(0)})/k$ and thus

(6) $\dim(A/P_0) + $ tr. deg. $\kappa(A)/k = $ tr. deg. $\kappa(P_0)/k$

since $\kappa(A/P_0) = \kappa(A)$ and $(A/P_0)_{(0)} = \kappa(P_0)$. (4) is now an immediate consequence of (6).

Hence it remains only to establish (3). Suppose first that (3) holds whenever A is an integral domain. Choose a minimal prime ideal P_0 of A such that $\dim A = \dim(A/P_0)$. Then (3) is an immediate consequence of (6) and (4).

Hence to establish (3) we may assume without loss of generality that A is an integral domain. We proceed by induction on $r = \dim A$. If $r = 0$ then A is a field and (3) holds trivially. Now let $r \geq 1$ and suppose (3) holds whenever $\dim A < r$. Let K denote the residue class field of A, L denote the quotient field of A and $\mathfrak{m}$ denote the maximal ideal of A.

Since $\dim A = r$ we have a chain $(0) \subsetneq P_1 \subsetneq \ldots \subsetneq P_r = \mathfrak{m}$ of prime ideals in A. Put $C = A/P_1$ and $L_1 = C_{(0)} = \kappa(P_1)$. Since $\dim C = r - 1$, by the inductive hypothesis we conclude

(7) $$\begin{cases} r - 1 + \text{tr. deg. } K/k = \dim C + \text{tr. deg. } K/k = \\ = \text{tr. deg. } L_1/k. \end{cases}$$

To establish (3) for A we need to show that $r + \text{tr. deg. } K/k = \text{tr. deg. } L/k$, or equivalently, in view of (7),

(8) $$\text{tr. deg. } L/k = \text{tr. deg. } L_1/k + 1.$$

Hence to establish (3) it remains only to establish (8). Since A is an integral domain, $A = B_Q$ where B is a finitely generated k-algebra, $Q \in \text{Spec}(B)$ and P_1 is a prime ideal in A of height 1, there exists $n \geq 1$ and $P, Q \in \text{Spec}(k[T_1, \ldots, T_n])$ such that $P \subset Q$, $A = (k[T_1, \ldots, T_n]/P)_Q$ and $P_1 = Q(k[T_1, \ldots, T_n]/P)_Q$. Hence L_1 is the residue class field of $k[T_1, \ldots, T_n]_Q$ and L is the residue class field of $k[T_1, \ldots, T_n]_P$. Thus (8) is the conclusion of Lemma A.5.2. This completes the proof of the theorem.

A.6.1. Corollary. Let k, B, Q and A be as in Thm. A.6. Then there exists $f \in B$ such that $Q \in D(f)$ and such that $\dim B_f = \dim A + \text{tr. deg. } K/k$, where K denotes the residue class field of A. Moreover, we can also choose f so that, in addition, $\text{tr. deg. } \kappa(P)/k = \dim B_f$ for all minimal prime ideals P in B_f.

Proof.

Note

(1) $$\text{Spec}(A) = \bigcap_{Q \in D(f),\, f \in B} \text{Spec}(B_f).$$

In view of (1) and conditions (2) and (3) in Thm. A.6 we conclude

(2) $$\begin{cases} \dim B_f \geq \dim A + \text{tr. deg. } K/k \text{ for all} \\ f \in B \text{ such that } Q \in D(f). \end{cases}$$

The set S of minimal prime ideals of B is finite since B is noetherian. Put $T = \{P \mid P \in S \text{ and tr. deg. } \kappa(P)/k \neq \dim A + \text{tr. deg. } K/k\}$. Let $T = \{P_1, \ldots, P_m\}$. For each $1 \leq i \leq m$ choose $f_i \in P_i \setminus Q$ and put $f = f_1 \cdots f_m$. Then $f \in (\bigcap_{i=1}^{m} P_i) \setminus Q$. Hence $T \cap \text{Spec}(B_f) = \phi$. Thus in view of conditions (2) and (3) in Thm. A.6 we conclude

(3) $$\dim B_f \leq \dim A + \text{tr. deg. } K/k.$$

Note that (2) and (3) yield the first conclusion of the corollary and by construction we have the second conclusion of the corollary.

A.7. Theorem. Let A be a ring, B be a finitely presented A-algebra with structure homomorphism $\lambda : A \to B$, let M be a finitely presented B-module, $Q \in \mathrm{Spec}\,(B)$ and put $P = \lambda^{-1}(Q)$. Then the following are true.

(1). Suppose M_Q is a flat A_P-module. Then there exists an open neighborhood U of Q in $\mathrm{Spec}\,(B)$ such that for each $Q_1 \in U$, putting $P_1 = \lambda^{-1}(Q_1)$, we have M_{Q_1} is a flat A_{P_1}-module.

(2). Suppose B_Q is a flat A_P-module. Put $U = \{\lambda^{-1}(R) \mid R \in \mathrm{Spec}\,(B)$ and B_R is a flat $A_{\lambda^{-1}(R)}$-module$\}$. Then U is a nonempty open subset of $\mathrm{Spec}\,(A)$.

Proof.

For a proof see Thm. 11.3.1, EGA IV, part 3 (reference [2]).

A.8. Proposition. Let A be a ring, B be an A-algebra with structure homomorphism $\lambda : A \to B$, let $Q \in \mathrm{Spec}\,(B)$, put $P = \lambda^{-1}(Q)$ and let $P_1 \in \mathrm{Spec}\,(A)$ such that $P_1 \subset P$. Suppose B_Q is a flat A_P-module. Then there exists $Q_1 \in \mathrm{Spec}\,(B)$ such that $Q_1 \subset Q$ and such that $P_1 = \lambda^{-1}(Q_1)$.

Proof.

This is a well-known result and is a special case of Prop. 3.9.3, p. 253 in [3].

A.9. Proposition. Let A be a noetherian ring, B be a finitely generated A-algebra with structure homomorphism $\lambda : A \to B$, let $Q \in \mathrm{Spec}\,(B)$ and put $P = \lambda^{-1}(Q)$. Suppose A_P is a regular local ring and B_Q is a Cohen-Macauley local ring. Then the following conditions are equivalent:

(1). B_Q is a flat A_P-module.

(2). $\dim(B_Q) = \dim(A_P) + \dim(B_Q \otimes_{A_P} \kappa(P))$.

Proof.

For a proof see 15.4.2, p. 230 in EGA IV, part 3 (reference [2]).

A.10. **Theorem.** (Criterion of flatness by fibers). Let A be a ring, let B be a finitely presented A-algebra with structure homomorphism $\lambda : A \to B$, let C be a B-algebra with structure homomorphism $\mu : B \to C$ which is a finitely presented A-algebra, let M be a finitely presented C-module, let $R \in \mathrm{Spec}(C)$, put $Q = \mu^{-1}(R)$ and $P = \lambda^{-1}(Q)$. Suppose $M_R \neq 0$. Then the following conditions are equivalent:

(1). M_R is a flat A_P-module and $M \otimes_A \kappa(P)$ is a flat $B \otimes_A \kappa(P)$-module.

(2). B_Q is a flat A_P-module and M_R is a flat B_Q-module.

Proof.

This is a well-known result and is a special case of Thm. 11.3.10, p. 138, EGA IV, part 3 (reference [2]).

A.11. **Proposition.** Let B be a ring, M be a finitely presented B-module and N be any B-module. Then for each $Q \in \mathrm{Spec}(B)$ there exists $f \in B$ such that $Q \in D(f)$ and such that the canonical homomorphism of B_Q-modules $(\mathrm{Hom}_{B_f}(M_f, N_f))_Q \to \mathrm{Hom}_{B_Q}(M_Q, N_Q)$ is bijective.

Proof.

For a proof see 4.1.1 in [1].

A.11.1. Corollary. Let B be a ring and M and N be finitely presented B-modules. Suppose there exists $Q \in \text{Spec}(B)$ such that M_Q and N_Q are isomorphic B_Q-modules. Then there exists $f \in B$ such that $Q \in D(f)$ and such that M_f and N_f are isomorphic B_f-modules.

Proof.

Indeed, let $h_1 : M_Q \to N_Q$ and $h_2 : N_Q \to M_Q$ be two inverse isomorphisms of B_Q-modules. By Prop. A.11 for each $i = 1, 2$ there exist $f_i \in B$ such that $Q \in D(f_i)$ and $g_i'' \in \text{Hom}_{B_{f_i}}(M_{f_i}, N_{f_i})$ such that $g_{i_Q}'' = h_i$. Put $f' = f_1 f_2$ and $g_i' = g_{i_{f'}}''$ for $i = 1, 2$. Then $g_{i_Q}' = h_i$ for $i = 1, 2$. Hence $(g_1' \circ g_2')_Q$ and $(g_2' \circ g_1')_Q$ being the identity automorphisms, applying Prop. A.11 again we obtain $f_0 \in B$ such that $Q \in D(f_0)$ and such that $(g_1' \circ g_2')_{f_0}$ and $(g_2' \circ g_1')_{f_0}$ are the identity automorphisms. Hence choosing $f = f' f_0$ suffices.

A.11.2. Corollary. Let B be a ring and M be a finitely presented B-module. Let $Q \in \text{Spec}(B)$ and n be an integer ≥ 1 such that M_Q is a free B_Q-module of rank n. Then there exists $f \in B$ such that $Q \in D(f)$ and such that M_f is a free B_f-module of rank n.

Proof.

Immediate by Cor. A.11.1.

A.12. Lemma. Let A be a ring, M be a finitely generated A-module, I be a set, B_i be a ring and $\lambda_i : A \to B_i$ be a ring homomorphism for all $i \in L$. Suppose $\bigcap_{i \in I} \text{Ker}(\lambda_i) = \{0\}$ and let n be the minimum cardinality of a set of generators for M as an A-module.

Then M is a locally free A-module of rank n if and only if $M \underset{A}{\otimes} B_i$ is a locally free B_i-module of rank n for each $i \in I$.

Proof.

For each $i \in I$ let $\mu_i : M \to M \underset{A}{\otimes} B_i$ be the canonical homomorphism and let $S = \{x_1, \ldots, x_n\}$ be a set of generators for M as an A-module. Suppose $M \underset{A}{\otimes} B_i$ is a locally free B_i-module of rank n for each $i \in I$ in order to prove M is a locally free A-module of rank n. This will establish the lemma since the converse is obvious. Suppose S is linearly dependent. Hence there exists $a_i \in A$ for each $1 \leq i \leq n$ such that $a_1 x_1 + \ldots + a_n x_n = 0$. Since S generates M as an A-module, $\{\mu_i(x_1), \ldots, \mu_i(x_n)\}$ generates $M \underset{A}{\otimes} B_i$ as a B_i-module for each $i \in I$ and thus is a basis. Since for each $i \in I$, $0 = \mu_i(a_1 x_1 + \ldots + a_n x_n) = \lambda_i(a_1)\mu_i(x_1) + \ldots + \lambda_i(a_n)\mu_n(x_n)$, we conclude that $\{a_1, \ldots, a_n\} \subset \underset{i \in I}{\bigcap} \text{Ker}(\lambda_i) = \{0\}$. This establishes the lemma.

A.12.1. Corollary. Let A be a reduced ring, M be a finitely generated A-module, let n be an integer ≥ 0 and suppose for each $P \in \text{Spec}(A)$ that n is the minimum cardinality of a set of generators for M_P as an A_P-module. Then M is a locally free A-module of rank n.

Proof.

By hypothesis and Nakayama's lemma we conclude that $M \underset{A}{\otimes} \kappa(P)$ is a $\kappa(P)$-vector space of dimension n for each $P \in \text{Spec}(A)$. Hence the corollary is immediate by Lemma A.12 since A is reduced, taking I to be the set of all prime ideals of A, $B_i = \kappa(P)$ for each $i = P \in I$ and $\lambda_i : A \to \kappa(P)$ to be the canonical homomorphism for each $i = P \in I$.

A.13. Proposition. Let A be a ring, B be a faithfully flat A-algebra, M be an A-module and put $N = M \underset{A}{\otimes} B$. Then M is a locally free A-module of finite type if and only if N is a locally free B-module of finite type.

Proof.

See Cor. 1.11, Chp. VIII, p. 201 in Séminaire de Géométrie Algébrique 1 (SGA1) by A. Grothendieck for a proof.

A.14. Lemma. Let A be a ring, M be a finitely generated A-module, $\{x_1, \ldots, x_n\} \subset M$, $Q \in \text{Spec}(A)$ and let $\lambda : M \to M_Q$ be the canonical homomorphism. Suppose $\{\lambda(x_1), \ldots, \lambda(x_n)\}$ generates the A_Q-module M_Q. Then there exists $f \in A$ such that $Q \in D(f)$ and such that letting $\mu : A \to A_f$ be the canonical homomorphism, $\{\mu(x_1), \ldots, \mu(x_n)\}$ generates the A_f-module M_f.

Proof.

This lemma is a special case of Prop. 5.2.2, p. 109, Chp. 0 in [3].

A.14.1. Corollary. Let A be a ring, M be a finitely generated A-module and suppose for each $P \in \text{Spec}(A)$ that M_P is a free A_P-module of finite rank r_P. Then M is a finitely presented A-module and for each $P \in \text{Spec}(A)$ there exists $f \in A$ such that $P \in D(f)$ and such that M_f is a free A_f-module of rank r_P.

Proof.

Immediate by Lemma A.14 and Thm. 1, Chap. II, §5, No. 2, p. 109 in Bourbaki, Commutative Algebra.

A.15. Proposition. Let k be a field, B be a finitely generated k-algebra and let $P \in \text{Spec}(B)$. Then P is a maximal ideal in B if and only if $\kappa(P)$ is a finite field extension of k.

Proof.

This is a well-known result.

A.16. Proposition. Let A be a ring, $Q \in \text{Spec}(A)$ and $a \in A$ such that the image of a in $\kappa(Q)$ under the canonical homomorphism $A \to \kappa(Q)$ is nonzero. Then there exists an open neighborhood U of Q in Spec(A) such that for each $P \in U$, the image of a in $\kappa(P)$ under the canonical homomorphism $A \to \kappa(P)$ is nonzero.

Proof.

This result is a special case of Prop. 5.5.1, Chap. 0, p. 119 in [3].

A.17. Proposition. Let k be a field, B be a finitely generated k-algebra, $Q \in \text{Spec}(B)$ and put $A = B_Q$. The following conditions are equivalent:

(1). For each purely inseparable algebraic field extension k' of k, the local ring $A \otimes_k k'$ is regular.

(2). For each finite purely inseparable algebraic field extension k' of k, the local ring $A \otimes_k k'$ is regular.

(3). Putting $k' = k^{p^{-\infty}}$, where p is the characteristic exponent of k, the local ring $A \otimes_k k'$ is regular.

(4). For each finite field extension k' of k, all the local rings of $A \otimes_k k'$ are regular.

(5). For each field extension k' of k such that k' is a finitely generated k-algebra, all the local rings of $A \otimes_k k'$ are regular.

(6). For each perfect field extension k' of k, all the local rings of $A \otimes_k k'$ are regular.

(7). For each field extension k' of k, where $k' = k^{p^{-s}}$, p is the characteristic exponent of k and s is an integer > 0, the local ring $A \otimes_k k'$ is regular.

(8). For each field extension k' of k, all the local rings of $A \otimes_k k'$ are regular.

Proof.

This is a well-known result. See Chp. IV, part 2 in [2], for example, for a proof.

A.17.1. Remark. A ring A with the properties in Prop. A.17 is said to be <u>geometrically regular</u>.

A.18. Theorem. With notation and hypotheses as in Thm. 9.2 let M_{i_0} be a finitely presented C_{i_0}-module, for each $i \in I$ such that $i \geq i_0$ put $M_i = M_{i_0} \otimes_{A_{i_0}} A_i$ yielding a direct system of C_{i_0}-modules $(M_i)_{i \in I, i \geq i_0}$ and put

$M = \varinjlim_{i \in I,\ i \geq i_0} M_i$ whence $M = M_{i_0} \otimes_{A_{i_0}} A$. The following are true:

(1). Let $Q \in \text{Spec}(C)$, put $P = \gamma^{-1}(Q)$ and for each $i \in I$, $i \geq i_0$ let Q_i be the preimage of Q in C_i and put $P_i = \gamma_i^{-1}(Q_i)$. Then M_Q is a flat B_P-module if and only if there exists $i \in I$, $i \geq i_0$ such that $M_{i_{Q_i}}$ is a flat $B_{i_{P_i}}$-module.

(2). M is a flat B-module if and only if there exists $i \in I$, $i \geq i_0$ such that M_i is a flat B_i-module.

Proof.

This is a well-known result and is a special case of Thm. 11.2.6, p. 123 in part 3 of Chp. IV in [2].

A.19. Theorem. Let A_0 be a ring, A be an A_0-algebra, B be a finitely presented A-algebra and let C be a finitely presented B-algebra with structure homomorphism $\gamma : B \to C$.

Then there exists a direct system $(A_i)_{i \in I}$ of finitely presented A_0-algebras (which can be taken to be subalgebras of A) indexed by a directed set I, an element i_0 of I and a finitely presented A_{i_0}-algebra B_{i_0} and a finitely presented B_{i_0}-algebra C_{i_0} with structure homomorphism $\gamma_{i_0} : B_{i_0} \to C_{i_0}$ such that putting $B_i = B_{i_0} \otimes_{A_{i_0}} A_i$ and $C_i = C_{i_0} \otimes_{A_{i_0}} A_i$ for each $i \in I$ such that $i \geq i_0$ we obtain direct systems $(B_i)_{i \in I,\ i \geq i_0}$ and $(C_i)_{i \in I,\ i \geq i_0}$ of A_{i_0}-algebras such that putting $\gamma_i = \gamma_{i_0} \otimes_{A_{i_0}} A_i : B_i \to C_i$ for each $i \in I$, $i \geq i_0$ the following are true: $A = \varinjlim_{i \in I} A_i$, $B = \varinjlim_{i \in I,\ i \geq i_0} B_i$, $C = \varinjlim_{i \in I,\ i \geq i_0} C_i$ and $\gamma = \varinjlim_{i \in I,\ i \geq i_0} \gamma_i$.

Proof.

This is a well-known result and is an immediate consequence of the well-known results in Chap. 0, §6 in [3].

A. 20. Lemma. Let A be a local ring, M be an A-module, n be an integer ≥ 1 and $\{x_1, \ldots, x_n\}$ and $\{y_1, \ldots, y_n\}$ be two subsets of M such that $Ax_1 + \ldots + Ax_n = Ay_1 + \ldots + Ay_n$.

Then $(y_1, \ldots, y_n)$ can be obtained from $(x_1, \ldots, x_n)$ by operations of the following types, letting $(z_1, \ldots, z_n) \in M^n$ be an n-tuple of elements of M obtained from $(x_1, \ldots, x_n)$ by operations of the following types:

Type 1. Interchanging z_1 and z_i for some $2 \leq i \leq n$.

Type 2. Replacing z_1 by az_1 for some unit $a \in A$.

Type 3. Replacing z_1 by $z_1 + az_2$ for some $a \in A$.

Proof.

See Lemma 1.9 in [6] for a proof.

A. 21. Lemma. Let A be a ring, B be a flat finitely presented A-algebra with structure homomorphism $\lambda : A \to B$, let I be an ideal in A contained in the nilradical of A, let $\lambda' : A/I \to B/IB$ be the homomorphism induced by λ and suppose that IB is a finitely generated ideal in B.

If λ' is bijective then so is λ.

Proof.

Since B is a finitely presented A-algebra, the ideal IB in B is finitely generated and λ' is an isomorphism, we conclude that B is a finitely generated A-module. Note that Spec (A/I) identifies to Spec (A) and Spec (B/IB) identifies to Spec (B) since I is contained in the nilradical of A and IB is contained in the nilradical of B. Since λ' is an isomorphism we thus conclude that the map Spec (B) → Spec (A) defined by $A \to \lambda^{-1}(Q)$ is surjective. Since λ is flat we thus conclude that λ is faithfully flat. Since λ is faithfully flat we conclude that λ is injective and $\lambda^{-1}(IB) = I$. The latter equality and the fact that λ' is bijective imply λ is surjective. This establishes the lemma.

BIBLIOGRAPHY

[1] A. Grothendieck, <u>Sur quelques points d'algèbre homologique</u>, Tohoku Math. Jour., t. IX (1957), pp. 119-221.

[2] A. Grothendieck, <u>Éléments de Géométrie Algébrique</u> (EGA), Institut des Hautes Études Scientifiques, Publ. Math.

[3] A. Grothendieck and J. Dieudonné, <u>Éléments de Géométrie Algébrique</u> (EGA), Springer-Verlag, 1971.

[4] H. Matsumura, <u>Commutative Algebra</u>, W. A. Benjamin Co., New York, 1970.

[5] M. Nagata, <u>Local Rings</u>, Interscience Tracts in Pure & Applied Math., 13, J. Wiley, New York, 1962.

[6] R. Sot, <u>Canonical classes in p-adic cohomology</u>, Univ. of Rochester, 1980 (thesis).

Index to Terminology

derivation, Def. 2.2

étale algebra, Def. 4.2.1

étale homomorphism, Def. 4.2.2

étale morphism, Def. 10.3

fitting ideal, Def. 6.1

generic point, Def. 1.1

geometrically regular, Rmk. 7.8.1

Jacobian criterion, Def. 1.3, Thm. 7.1

Kahler differentials, Def. 2.1, Rmk. 10.3.1

Kahler differentials, sheaf of, Rmk. 10.3.1

prime spectrum, Def. 1.1

simple algebra, Def. 1.3, Def. 3.3, Def. 7.2

simple homomorphism, Def. 7.2

simple morphism, Def. 10.2

unramified algebra, Def. 4.2.2

unramified homomorphism, Def. 4.2.2

unramified morphism, Def. 10.3

Zariski topology, Def. 1.1

Index to Symbols

Spec (A), Def. 1.1

$D(f)$, Def. 1.1

$\kappa(P)$, 1.2

$\alpha(P)$, 1.2

$I_{B/A}$, I_B, I, Def. 2.1

$\Gamma_A^1(B)$, Def. 2.1

$d_{B/A}$, d, Def. 2.1

$\text{Der}_A(B,M)$, Rmk. 2.2.2

$D_B(M)$, Def. 2.5

$I_r^{A,(*)}(M)$, I_r, Def. 6.1

$\Gamma_Y^1(X)$, Rmk. 10.3.1

Vol. 787: Potential Theory, Copenhagen 1979. Proceedings, 1979. Edited by C. Berg, G. Forst and B. Fuglede. VIII, 319 pages. 1980.

Vol. 788: Topology Symposium, Siegen 1979. Proceedings, 1979. Edited by U. Koschorke and W. D. Neumann. VIII, 495 pages. 1980.

Vol. 789: J. E. Humphreys, Arithmetic Groups. VII, 158 pages. 1980.

Vol. 790: W. Dicks, Groups, Trees and Projective Modules. IX, 127 pages. 1980.

Vol. 791: K. W. Bauer and S. Ruscheweyh, Differential Operators for Partial Differential Equations and Function Theoretic Applications. V, 258 pages. 1980.

Vol. 792: Geometry and Differential Geometry. Proceedings, 1979. Edited by R. Artzy and I. Vaisman. VI, 443 pages. 1980.

Vol. 793: J. Renault, A Groupoid Approach to C*-Algebras. III, 160 pages. 1980.

Vol. 794: Measure Theory, Oberwolfach 1979. Proceedings 1979. Edited by D. Kölzow. XV, 573 pages. 1980.

Vol. 795: Séminaire d'Algèbre Paul Dubreil et Marie-Paule Malliavin. Proceedings 1979. Edited by M. P. Malliavin. V, 433 pages. 1980.

Vol. 796: C. Constantinescu, Duality in Measure Theory. IV, 197 pages. 1980.

Vol. 797: S. Mäki, The Determination of Units in Real Cyclic Sextic Fields. III, 198 pages. 1980.

Vol. 798: Analytic Functions, Kozubnik 1979. Proceedings. Edited by J. Ławrynowicz. X, 476 pages. 1980.

Vol. 799: Functional Differential Equations and Bifurcation. Proceedings 1979. Edited by A. F. Izé. XXII, 409 pages. 1980.

Vol. 800: M.-F. Vignéras, Arithmétique des Algèbres de Quaternions. VII, 169 pages. 1980.

Vol. 801: K. Floret, Weakly Compact Sets. VII, 123 pages. 1980.

Vol. 802: J. Bair, R. Fourneau, Etude Géometrique des Espaces Vectoriels II. VII, 283 pages. 1980.

Vol. 803: F.-Y. Maeda, Dirichlet Integrals on Harmonic Spaces. X, 180 pages. 1980.

Vol. 804: M. Matsuda, First Order Algebraic Differential Equations. VII, 111 pages. 1980.

Vol. 805: O. Kowalski, Generalized Symmetric Spaces. XII, 187 pages. 1980.

Vol. 806: Burnside Groups. Proceedings, 1977. Edited by J. L. Mennicke. V, 274 pages. 1980.

Vol. 807: Fonctions de Plusieurs Variables Complexes IV. Proceedings, 1979. Edited by F. Norguet. IX, 198 pages. 1980.

Vol. 808: G. Maury et J. Raynaud, Ordres Maximaux au Sens de K. Asano. VIII, 192 pages. 1980.

Vol. 809: I. Gumowski and Ch. Mira, Recurences and Discrete Dynamic Systems. VI, 272 pages. 1980.

Vol. 810: Geometrical Approaches to Differential Equations. Proceedings 1979. Edited by R. Martini. VII, 339 pages. 1980.

Vol. 811: D. Normann, Recursion on the Countable Functionals. VIII, 191 pages. 1980.

Vol. 812: Y. Namikawa, Toroidal Compactification of Siegel Spaces. VIII, 162 pages. 1980.

Vol. 813: A. Campillo, Algebroid Curves in Positive Characteristic. V, 168 pages. 1980.

Vol. 814: Séminaire de Théorie du Potentiel, Paris, No. 5. Proceedings. Edited by F. Hirsch et G. Mokobodzki. IV, 239 pages. 1980.

Vol. 815: P. J. Slodowy, Simple Singularities and Simple Algebraic Groups. XI, 175 pages. 1980.

Vol. 816: L. Stoica, Local Operators and Markov Processes. VIII, 104 pages. 1980.

Vol. 817: L. Gerritzen, M. van der Put, Schottky Groups and Mumford Curves. VIII, 317 pages. 1980.

Vol. 818: S. Montgomery, Fixed Rings of Finite Automorphism Groups of Associative Rings. VII, 126 pages. 1980.

Vol. 819: Global Theory of Dynamical Systems. Proceedings, 1979. Edited by Z. Nitecki and C. Robinson. IX, 499 pages. 1980.

Vol. 820: W. Abikoff, The Real Analytic Theory of Teichmüller Space. VII, 144 pages. 1980.

Vol. 821: Statistique non Paramétrique Asymptotique. Proceedings, 1979. Edited by J.-P. Raoult. VII, 175 pages. 1980.

Vol. 822: Séminaire Pierre Lelong–Henri Skoda, (Analyse) Années 1978/79. Proceedings. Edited by P. Lelong et H. Skoda. VIII, 356 pages, 1980.

Vol. 823: J. Král, Integral Operators in Potential Theory. III, 171 pages. 1980.

Vol. 824: D. Frank Hsu, Cyclic Neofields and Combinatorial Designs. VI, 230 pages. 1980.

Vol. 825: Ring Theory, Antwerp 1980. Proceedings. Edited by F. van Oystaeyen. VII, 209 pages. 1980.

Vol. 826: Ph. G. Ciarlet et P. Rabier, Les Equations de von Kármán. VI, 181 pages. 1980.

Vol. 827: Ordinary and Partial Differential Equations. Proceedings, 1978. Edited by W. N. Everitt. XVI, 271 pages. 1980.

Vol. 828: Probability Theory on Vector Spaces II. Proceedings, 1979. Edited by A. Weron. XIII, 324 pages. 1980.

Vol. 829: Combinatorial Mathematics VII. Proceedings, 1979. Edited by R. W. Robinson et al.. X, 256 pages. 1980.

Vol. 830: J. A. Green, Polynomial Representations of GL_n. VI, 118 pages. 1980.

Vol. 831: Representation Theory I. Proceedings, 1979. Edited by V. Dlab and P. Gabriel. XIV, 373 pages. 1980.

Vol. 832: Representation Theory II. Proceedings, 1979. Edited by V. Dlab and P. Gabriel. XIV, 673 pages. 1980.

Vol. 833: Th. Jeulin, Semi-Martingales et Grossissement d'une Filtration. IX, 142 Seiten. 1980.

Vol. 834: Model Theory of Algebra and Arithmetic. Proceedings, 1979. Edited by L. Pacholski, J. Wierzejewski, and A. J. Wilkie. VI, 410 pages. 1980.

Vol. 835: H. Zieschang, E. Vogt and H.-D. Coldewey, Surfaces and Planar Discontinuous Groups. X, 334 pages. 1980.

Vol. 836: Differential Geometrical Methods in Mathematical Physics. Proceedings, 1979. Edited by P. L. García, A. Pérez-Rendón, and J. M. Souriau. XII, 538 pages. 1980.

Vol. 837: J. Meixner, F. W. Schäfke and G. Wolf, Mathieu Functions and Spheroidal Functions and their Mathematical Foundations Further Studies. VII, 126 pages. 1980.

Vol. 838: Global Differential Geometry and Global Analysis. Proceedings 1979. Edited by D. Ferus et al. XI, 299 pages. 1981.

Vol. 839: Cabal Seminar 77 – 79. Proceedings. Edited by A. S. Kechris, D. A. Martin and Y. N. Moschovakis. V, 274 pages. 1981.

Vol. 840: D. Henry, Geometric Theory of Semilinear Parabolic Equations. IV, 348 pages. 1981.

Vol. 841: A. Haraux, Nonlinear Evolution Equations- Global Behaviour of Solutions. XII, 313 pages. 1981.

Vol. 842: Séminaire Bourbaki vol. 1979/80. Exposés 543–560. IV, 317 pages. 1981.

Vol. 843: Functional Analysis, Holomorphy, and Approximation Theory. Proceedings. Edited by S. Machado. VI, 636 pages. 1981.

Vol. 844: Groupe de Brauer. Proceedings. Edited by M. Kervaire and M. Ojanguren. VII, 274 pages. 1981.

Vol. 845: A. Tannenbaum, Invariance and System Theory: Algebraic and Geometric Aspects. X, 161 pages. 1981.

Vol. 846: Ordinary and Partial Differential Equations, Proceedings. Edited by W. N. Everitt and B. D. Sleeman. XIV, 384 pages. 1981.

Vol. 847: U. Koschorke, Vector Fields and Other Vector Bundle Morphisms – A Singularity Approach. IV, 304 pages. 1981.

Vol. 848: Algebra, Carbondale 1980. Proceedings. Ed. by R. K. Amayo. VI, 298 pages. 1981.

Vol. 849: P. Major, Multiple Wiener-Itô Integrals. VII, 127 pages. 1981.

Vol. 850: Séminaire de Probabilités XV. 1979/80. Avec table générale des exposés de 1966/67 à 1978/79. Edited by J. Azéma and M. Yor. IV, 704 pages. 1981.

Vol. 851: Stochastic Integrals. Proceedings, 1980. Edited by D. Williams. IX, 540 pages. 1981.

Vol. 852: L. Schwartz, Geometry and Probability in Banach Spaces. X, 101 pages. 1981.

Vol. 853: N. Boboc, G. Bucur, A. Cornea, Order and Convexity in Potential Theory: H-Cones. IV, 286 pages. 1981.

Vol. 854: Algebraic K-Theory. Evanston 1980. Proceedings. Edited by E. M. Friedlander and M. R. Stein. V, 517 pages. 1981.

Vol. 855: Semigroups. Proceedings 1978. Edited by H. Jürgensen, M. Petrich and H. J. Weinert. V, 221 pages. 1981.

Vol. 856: R. Lascar, Propagation des Singularités des Solutions d'Equations Pseudo-Différentielles à Caractéristiques de Multiplicités Variables. VIII, 237 pages. 1981.

Vol. 857: M. Miyanishi. Non-complete Algebraic Surfaces. XVIII, 244 pages. 1981.

Vol. 858: E. A. Coddington, H. S. V. de Snoo: Regular Boundary Value Problems Associated with Pairs of Ordinary Differential Expressions. V, 225 pages. 1981.

Vol. 859: Logic Year 1979–80. Proceedings. Edited by M. Lerman, J. Schmerl and R. Soare. VIII, 326 pages. 1981.

Vol. 860: Probability in Banach Spaces III. Proceedings, 1980. Edited by A. Beck. VI, 329 pages. 1981.

Vol. 861: Analytical Methods in Probability Theory. Proceedings 1980. Edited by D. Dugué, E. Lukacs, V. K. Rohatgi. X, 183 pages. 1981.

Vol. 862: Algebraic Geometry. Proceedings 1980. Edited by A. Libgober and P. Wagreich. V, 281 pages. 1981.

Vol. 863: Processus Aléatoires à Deux Indices. Proceedings, 1980. Edited by H. Korezlioglu, G. Mazziotto and J. Szpirglas. V, 274 pages. 1981.

Vol. 864: Complex Analysis and Spectral Theory. Proceedings, 1979/80. Edited by V. P. Havin and N. K. Nikol'skii, VI, 480 pages. 1981.

Vol. 865: R. W. Bruggeman, Fourier Coefficients of Automorphic Forms. III, 201 pages. 1981.

Vol. 866: J.-M. Bismut, Mécanique Aléatoire. XVI, 563 pages. 1981.

Vol. 867: Séminaire d'Algèbre Paul Dubreil et Marie-Paule Malliavin. Proceedings, 1980. Edited by M.-P. Malliavin. V, 476 pages. 1981.

Vol. 868: Surfaces Algébriques. Proceedings 1976-78. Edited by J. Giraud, L. Illusie et M. Raynaud. V, 314 pages. 1981.

Vol. 869: A. V. Zelevinsky, Representations of Finite Classical Groups. IV, 184 pages. 1981.

Vol. 870: Shape Theory and Geometric Topology. Proceedings, 1981. Edited by S. Mardešić and J. Segal. V, 265 pages. 1981.

Vol. 871: Continuous Lattices. Proceedings, 1979. Edited by B. Banaschewski and R.-E. Hoffmann. X, 413 pages. 1981.

Vol. 872: Set Theory and Model Theory. Proceedings, 1979. Edited by R. B. Jensen and A. Prestel. V, 174 pages. 1981.

Vol. 873: Constructive Mathematics, Proceedings, 1980. Edited by F. Richman. VII, 347 pages. 1981.

Vol. 874: Abelian Group Theory. Proceedings, 1981. Edited by R. Göbel and E. Walker. XXI, 447 pages. 1981.

Vol. 875: H. Zieschang, Finite Groups of Mapping Classes of Surfaces. VIII, 340 pages. 1981.

Vol. 876: J. P. Bickel, N. El Karoui and M. Yor. Ecole d'Eté de Probabilités de Saint-Flour IX – 1979. Edited by P. L. Hennequin. XI, 280 pages. 1981.

Vol. 877: J. Erven, B.-J. Falkowski, Low Order Cohomology and Applications. VI, 126 pages. 1981.

Vol. 878: Numerical Solution of Nonlinear Equations. Proceedings, 1980. Edited by E. L. Allgower, K. Glashoff, and H.-O. Peitgen. XIV, 440 pages. 1981.

Vol. 879: V. V. Sazonov, Normal Approximation – Some Recent Advances. VII, 105 pages. 1981.

Vol. 880: Non Commutative Harmonic Analysis and Lie Groups. Proceedings, 1980. Edited by J. Carmona and M. Vergne. IV, 553 pages. 1981.

Vol. 881: R. Lutz, M. Goze, Nonstandard Analysis. XIV, 261 pages. 1981.

Vol. 882: Integral Representations and Applications. Proceedings, 1980. Edited by K. Roggenkamp. XII, 479 pages. 1981.

Vol. 883: Cylindric Set Algebras. By L. Henkin, J. D. Monk, A. Tarski, H. Andréka, and I. Németi. VII, 323 pages. 1981.

Vol. 884: Combinatorial Mathematics VIII. Proceedings, 1980. Edited by K. L. McAvaney. XIII, 359 pages. 1981.

Vol. 885: Combinatorics and Graph Theory. Edited by S. B. Rao. Proceedings, 1980. VII, 500 pages. 1981.

Vol. 886: Fixed Point Theory. Proceedings, 1980. Edited by E. Fadell and G. Fournier. XII, 511 pages. 1981.

Vol. 887: F. van Oystaeyen, A. Verschoren, Non-commutative Algebraic Geometry, VI, 404 pages. 1981.

Vol. 888: Padé Approximation and its Applications. Proceedings, 1980. Edited by M. G. de Bruin and H. van Rossum. VI, 383 pages. 1981.

Vol. 889: J. Bourgain, New Classes of $\mathcal{L}^p$-Spaces. V, 143 pages. 1981.

Vol. 890: Model Theory and Arithmetic. Proceedings, 1979/80. Edited by C. Berline, K. McAloon, and J.-P. Ressayre. VI, 306 pages. 1981.

Vol. 891: Logic Symposia, Hakone, 1979, 1980. Proceedings, 1979, 1980. Edited by G. H. Müller, G. Takeuti, and T. Tugué. XI, 394 pages. 1981.

Vol. 892: H. Cajar, Billingsley Dimension in Probability Spaces. III, 106 pages. 1981.

Vol. 893: Geometries and Groups. Proceedings. Edited by M. Aigner and D. Jungnickel. X, 250 pages. 1981.

Vol. 894: Geometry Symposium. Utrecht 1980, Proceedings. Edited by E. Looijenga, D. Siersma, and F. Takens. V, 153 pages. 1981.

Vol. 895: J.A. Hillman, Alexander Ideals of Links. V, 178 pages. 1981.

Vol. 896: B. Angéniol, Familles de Cycles Algébriques – Schéma de Chow. VI, 140 pages. 1981.

Vol. 897: W. Buchholz, S. Feferman, W. Pohlers, W. Sieg, Iterated Inductive Definitions and Subsystems of Analysis: Recent Proof-Theoretical Studies. V, 383 pages. 1981.

Vol. 898: Dynamical Systems and Turbulence, Warwick, 1980. Proceedings. Edited by D. Rand and L.-S. Young. VI, 390 pages. 1981.

Vol. 899: Analytic Number Theory. Proceedings, 1980. Edited by M.I. Knopp. X, 478 pages. 1981.

MIX
Papier aus verantwortungsvollen Quellen
Paper from responsible sources
FSC® C105338

If you have any concerns about our products,
you can contact us on
ProductSafety@springernature.com

In case Publisher is established outside the EU,
the EU authorized representative is:
Springer Nature Customer Service Center GmbH
Europaplatz 3, 69115 Heidelberg, Germany

Printed by Libri Plureos GmbH
in Hamburg, Germany